AN ACCOUNT

OF THE

OPERATIONS IN BURMA

CARRIED OUT BY

PROBYN'S HORSE

DURING

FEBRUARY, MARCH & APRIL 1945

EDITED BY

MAJOR B. H. MYLNE, M.B.E.

AN ACCOUNT OF THE

OPERATIONS IN BURMA

CARRIED OUT BY

PROBYN'S HORSE

DURING

FEBRUARY, MARCH & APRIL 1945

EDITED BY
MAJOR B. H. MYLNE

The Naval & Military Press Ltd

Published by
The Naval & Military Press Ltd
Unit 10, Ridgewood Industrial Park,
Uckfield, East Sussex,
TN22 5QE England
Tel: +44 (0) 1825 749494
Fax: +44 (0) 1825 765701
www.naval-military-press.com

In reprinting in facsimile from the original, any imperfections are inevitably reproduced and the quality may fall short of modern type and cartographic standards.

EDITOR'S NOTE

At *Meiktila* it was decided that notes written by the Colonel and Squadron Commanders after individual actions should be collated for the information of both officers absent from the Regiment and of old members of the Regiment. It was intended to produce these notes in three parts :—

Part 1. The advance to and capture of *Meiktila.*
Part 2. The advance from *Meiktila* to *Rangoon.*
Part 3. The story of any subsequent operations.

It was also intended that these three parts should be rewritten enventually by one who specialises in the compiling of regimental histories.

Part 1 was prepared in B3 Echelon, with difficulty while the Regiment was on the move, and sent out from *Pegu.* On arrival in *Rangoon,* it was decided that Part 1 should be rewritten and combined with Part 2 and that both parts should be printed together with sketch maps, and photographs. This has been done, but the printing has taken far longer than was estimated. Part 3 will never now be written.

Any comments, additions or suggestions which should be included in the "Official Account of the Part Played by Probyns Horse during the Japanese Campaign" when it is written will be more than welcome, both from those who were present and those who were not.

Oct 45. B.H.M.
Rangoon.

A SHERMAN TANK CROSSING THE IRRAWADDY

CONTENTS.

THE IRRAWADDY RIVER

CHAPTER I.

255 Indian Tank Brigade, consisting of 116 R. A. C. (Gordon Highlanders), Probyn's Horse, and the Royal Deccan Horse concentrated in *Kanglatongbi* in *Assam* by the end of October 1944. Training took place during November, and exercises were carried out teaching lessons learnt by 254 Indian Tank Brigade during the operations around *Imphal* earlier in the year.

The move forward from *Kanglatongbi* was carried out on the axis *Palel - Tamu - Kalewa - Kan - Gangaw - Pauk.* The road in many places was hardly fit for transporters, and the surface was very rough causing undue wear and breakages. South of *Kan* the road was quite unfit for transporters.

4 Corps decided that one tank regiment should be made available to support 7 Indian Division in their task of advancing to and forming a bridgehead over the *Irrawaddy.* This task fell eventually to 116 R. A. C. (Gordon Highlanders) who happened to be ahead at the time. This regiment was put under command of 7 Indian Division with whom it operated for several weeks.

The actual assault across the *Irrawaddy* took place on the 14th February, 1945 and did not achieve local surprise. Many of the assault boats were swept by machine gun fire and failed to get ashore. For the first few hours the operation looked highly unpromising, and only a thin trickle of traffic was being landed. Then events took a more favourable turn as bridging equipment began to arrive in quantity, and enemy resistance was gradually overcome.

The crossing of all formations was controlled by 4 Corps, and a special route was allotted to 255 Indian Tank Brigade. The control organization of this route was under command of the brigade 2nd in command, (Col R. Younger DSO MC). Probyn's Horse was called upon to provide Major F. W. Kennedy, six officers, and the Intercommunication Troop. This party controlled the crossing of the brigade onto the South bank.

A force known as '*Tomcol*' commanded by Major D. H. Mudie the Royal Deccan Horse, the Scout Troops of Probyn's and the Royal Deccan Horse, and the truck company 6/7 Rajputs, had been formed with the object of passing through the bridgehead and establishing patrol bases. From these bases reconnaissances were ordered to discover the presence or otherwise of enemy in the *Ngathayauk*, *Welaung*, *Seiktein*, and *Taungtha* areas, and to report on the state of the roads. *Tomcol* was later reinforced by one squadron 11 Cavalry (P. A. V. O.) and command was taken over by Major Chaplin of that regiment.

The thrust from the *Irrawaddy* bridgehead to *Meiktila* was carried out on the axis *Nyaungu - Nagathayauk*, thence on a two brigade front, main axis *Kamye -Taungtha*, subsidiary axis *Seiktein - Welaung - Taungtha*. From *Taungtha* the whole force moved on the road *Meilang - Sedaw - Meiktila*. The road mileage from *Kanglatongbi* to *Meiktila* was approximately 470, of which 260 were covered on transporters.

The brigade had moved into an assembly area at *Aingye*, North West of *Myitche*, and started crossing on the afternoon of the 17th. B2 and B3 echelons and administrative units remained with 17 Indian Division' administrative base. Probyn's Horse crossed the River without incident on the 18th and 19th February, and remained in a concentration area on the 20th, preparatory to the advance on *Nagathayauk*.

Tomcol had met resistance at *Sewa*, and, after an airstrike, the enemy fled leaving behind a gun and an ammunition dump. It was thought that he would return during the night, as *Tomcol* without infantry was unable to hold *Sewa*. Briefly the plan for the capture of the *Nagathayauk* group of villages was as follows:—

1. One squadron of the Royal Deccan Horse with 48 Indian Infantry Brigade was to move along the main divisional axis of the road towards *Sewa* preceeded by *Tomcol*.

2. The remainder of 255 Indian Tank Brigade was to advance via *Wetlu*, South of the main road, to a point South West of *Sewa* from which the Royal Deccan Horse were to attack *Sewa*.

3. On the capture of *Sewa*, Probyn's Horse with the 6/7 Rajputs were to move across country and attack *Nagathayauk* from the South. Mopping up was the responsibility of 48 Indian Infantry Brigade.

Having arrived at *Wetlu* it was reported that *Sewa* was not occupied, and Probyn's Horse was therefore pushed straight forward to clear its objective. The Regiment moved due East from *Wellu* to the *Kyaukpadaung* road across difficult country where the absence of the Scout Troop, which was still with *Tomcol*, was found to be a handicap. It then formed up astride the road preparatory to advancing due North on to the objective. From air photographs it had appeared that the villages to be cleared, seven in all, were not extensive and that cover was light. During the move up to the start line, a nullah just South of the group of villages, some small arms fire was encountered from what were subsequently found to have been enemy withdrawing down the *Seiktein* road. The group of villages was found to be empty, but one Japanese N. C. O. was killed. The villages contained thick trees and hedges, visibility being limited to 50 yards. On entering the villages it was found that control was extremely difficult and made more so by the fact that a certain amount of prophylactic fire and grenading by the infantry set fire to the bashas. Although it was thought that the bounds selected from air photographs were small, these would actually have been far too extensive on the ground had the area been held. After clearing the villages, the Regiment moved into a brigade harbour South of *Ngathayauk*.

Early on the morning of the 22nd February, the Regiment was ordered to be ready to move at 1100 hours, with the 6/7 Rajputs and 59/18 Battery R. A. in support, to clear up enemy resistance at *Oyin* on the *Seiktein* road. The Regiment moved across country to M. S. 19 and formed up on the main road in column of squadrons, 'B' Squadron (Major Stewart) leading. An armoured car patrol was met at M. S. 19 which reported enemy snipers to be in the palm trees North of *Oyin*. The infantry were dismounted from the tanks of the leading squadrons, and the advance was continued on both sides of the road. Snipers were soon spotted in the palm trees and bushes and several were killed. The village was found to be held, and the North and West edges were engaged initially. There were numerous fox holes and some bunkers in the village itself, in the hedgerows on the side of the road and in the hedges in a field to the West of the village. The advance was carried out methodically by the leading squadron and a company of infantry, enemy resistance being destroyed where ever it was encountered. It was found, however, that several positions on the side of the road had been missed, and so No 2 Troop of 'C' Squadron (Ris Mohd Arif) was called up and moved 'line ahead' down the road behind 'B' Squadron. The troop dealt with these positions.

Numerous assaults were made onto the tanks by individual Japs with picric acid boxes, but these were usually dealt with by the infantry or by the tanks themselves before any damage could be done. Only one tank, a squadron commander's (Major Stewart), was put out of action by a Jap who got beneath the tank and blew up himself and the tank, caving in the bottom plate, smashing in the gear box, and killing the driver, Swr Dayal Singh. One Jap got onto the turret of the second tank (Jem Mohd Shaffi) of the 'C' Squadron troop moving down the road, but was unable to open the turret and throw in his grenade. The tank commander called '***Budmash***', the signal for tank hunters, on his 'B' set, and the troop leader's tank traversed and shot off the Jap. At the same time a second Jap, with a picric acid charge, got under the same tank. The troop leader ordered the tank to reverse, thereby exposing the Jap, and killed him too. The Jap had omitted to pull the detonator of his charge and the tank was undamaged. A third assault on Jem Feroze Khan's tank was made by a Jap who attempted to climb onto the front of the tank carrying a mine. His body got caught in the track or sprocket and landed on the back of the tank, the mine having detonated without damage. 'B' Squadron rallied at the South end of the village while the infantry continued to mop up the West half.

'C' Squadron (Major Arkinstall) was then called forward to clear up the East half of the village. The squadron engaged several Japs moving out of the village to the East, destroyed one or two positions, but met no real resistance. 'B' Squadron was then ordered to consolidate on the North and West faces, 'C' Squadron on the South and East. It was now getting late and sniping and small arms fire were still reported from the jheel and tamarin trees on the East side of the village. 'A' Squadron (Major Loraine-Smith) was therefore ordered to sweep round the East face and to clear up the jheel area before moving into harbour just South of the village. Between the village and the jheel were numerous trees and some hedges. Round the jheel itself was a high bund and inside was a dense mass of trees. 'A' Squadron moved 'two up' between the jheel and the village and was soon hotly engaged from a bunker on the bank of the jheel, a hut in the trees near the village, and numerous snipers in the trees. The infantry were pinned to the ground and were unable to indicate targets. Both bunkers were destroyed, but heavy sniping and LMG fire continued from the trees around the jheel. One tank was sent along the North bank of the bund, but found no bunkers. Another tank got onto the bund, but could not get into the

MAJOR H. I. E. R. O. STEWART AND 'BOB'

jheel owing to the thickness of the cover. The right troop (Lieut Bahadur Singh) continued to advance, the troop leader's tank receiving forty hits on the turret from small arms fire and had three periscopes broken.

It was now dusk, and it was realised that unless the infantry advanced quickly through the opposition, the tanks and infantry were in danger of being caught out after dark. The Squadron was therefore ordered to move forward by very short bounds, bringing the infantry with them. The company commander with 'A' Squadron had been killed, and the infantry had become somewhat disorganized, and it was difficult to get some of them to advance. Major Loraine-Smith dismounted several times from his tank in order to encourage them. At the South edge of the village some Japs with a dog were seen crawling down a sunken road; they were engaged. On crossing this road the tank of the squadron 2nd in command (Capt. Nicolson) was assaulted by a tank hunter, and although a mine was placed against a track and exploded, no damage was done except that the camouflage net caught fire; the crew dismounted and extinguished the fire. Meanwhile the tank behind stuck on the road. Its commander Dfr Durga Singh put out his head and was killed by a sniper. The squadron then rallied with as many infantry as it had been able to bring forward, and moved into harbour. A tank was sent back to recover the tank which had stuck on the sunken road. Several explosions had taken place near this tank, and it was believed that Japs were blowing off their charges but lacked the courage to place them on the tanks.

The Regiment harboured South of the village without further incident, and no Japs were found in the area the following morning. Although the bund position was just as strongly defended as the rest of the village, it is believed that, had the infantry continued to advance boldly with the tanks, they would have suffered far fewer casualties. The chief difficulty from the tanks' point of view was in locating the opposition in the thick tamarin trees without the assistance of the infantry. The layout of the enemy defences appeared to be standard:—a screen of snipers in front, foxholes, bunkers, snipers in trees around the main position, and two battalion guns in rear, both of which were captured.

A summary of our own and enemy casualties is as follows:—

Own troops (including 6/7 Rajputs)

1	B. O. killed,	3	wounded.
1	V. C. O. killed	1	wounded.
15	I. O. Rs killed	37	wounded.
17		41	

Enemy

200 killed minimum, but possibly as many as 300. 2 Bn guns and many L.M.Gs. were captured and destroyed. The villagers stated later that the enemy had admitted 60 killed, but claimed that he had killed 300. An enemy ammunition dump was destroyed.

One tank which had been knocked out was later recovered by 116 R. A. C. (Gordon Highlanders).

The next phase of the advance was the move to and capture of *Taungtha.* The advance was made on two axes. Probyn's Horse, 59/18 Battery R.A., 6/7 Rajputs and 63 Indian Infantry Brigade, who had moved to *Oyin* after its capture, advanced via *Seiktein* and *Welaung.* The major part of 255 Indian Tank Brigade's B1 echelon moved behind 63 Indian Infantry Brigade; it included a large P. O. L. and ammunition column.

The Regiment moved to *Eywa* on the 23rd February, 'B' Squadron (Major Stewart) acting as advance guard under command of the 7/10 Baluch. This squadron had one small engagement, but the advance was not carried out as quickly as it might have been, had the advance guard been commanded by the tank commander, owing to the infantry's bad communications.

On the 24th February, the Regiment remained under command of 63 Indian Infantry Brigade. 'C' Squadron (Major Arkinstall) was under command of the Border Regiment, and formed part of the advance guard. Opposition was encountered in the *Sindewa Chaung,* and 'C' squadron quickly deployed on each side of the road. On reaching the chaung, the point troop (Capt Anderson) gave fire support to No 1 Troop (Capt Riszul Karim Khan) and to No 3 Troop (Ris Mohd Arif) which moved under the squadron 2nd in command (Capt O'Beirne Ryan) across the chaung on both sides of the road. The enemy offered no real resistance as they were not dug in, and withdrew rapidly leaving forty dead behind. We suffered no casualties. This was a much rehearsed vanguard action which worked like clock-work. No further opposition was encountered. The Regiment moved to harbour at M. S. 40 on the *Mahlaing* road and reverted to command of 255 Indian Tank Brigade. One troop (Lieut Grover) remained under command of 63 Indian Infantry Brigade to assist in the clearing up of small pockets of resistance in the area of M. S. 8. *Taungtha* had been captured during the day by 48 Indian Infantry Brigade with the Royal Deccan Horse under command.

The Brigade B1 Echelon, about 230 vehicles under command of Major B. H. Mylne, had been strafed by enemy planes during the afternoon. Several vehicles were destroyed and seventeen other ranks were wounded. 63 Indian Infantry Brigade did not reach its harbour until after dark and the road behind it and in front of the B1 Echelon was jammed by artillery vehicles. Reports were received that the rearguard was being followed up by enemy patrols. The brigade commander ordered Major Mylne to harbour on the roadside where he was, but Major Mylne refused on the grounds that he required a minimum of two companies with which to defend so large a number of vehicles. Eventually the brigade commander agreed to release a company and Lieut Grover's troop of tanks to reinforce the echelon's original escort company of the 6/7 Rajputs. At 2030 hours the harbour was bombed and strafed by a Jap plane, but to everyone's relief the bombs missed their target and no damage was done.

The next phase of the advance was the capture of *Mahlaing* and the *Thabutkon* Airfield. The Royal Deccan Horse, 4/4 Bombay Grenadiers, one squadron 11 Cavalry (P. A. V. O.), with 59/18 Battery R.A. in support, formed the advance guard of 255 Indian Tank Brigade. Probyns Horse moved in the main body. The Brigade harboured just North of *Mahlaing*..

On the 26th February, the Regiment with 6/7 Rajputs moved forward and cleared *Malhaing* without opposition. The operation served as an excellent rehearsal for future town fighting. The Brigade harboured at M. S. 17. The Royal Deccan Horse with supporting troops had done a wide left hook over difficult country via *Kwawgaung* and then captured

Thabutkon Airfield where a large consignment of petrol was landed.

The last phase of the advance was the move to *Meiktila.* On the 27th February, one squadron 16th Cavalry was sent forward with the task of recceing the approaches to *Meiktila.* One squadron of Probyn's Horse was sent to the *Thabutkon* Airfield with the object of providing a protective screen for the landing of 99 Indian Infantry Brigade who were scheduled to arrive by air. The armoured car squadron (16th Cavalry) ran into an enemy position at M. S. 8½ and was unable to get on. R. H. Q. and 'C' Squadron (Major Arkinstall) of Probyn's Horse were ordered up to join 63 Indian Infantry Brigade. At 1100 hours 'A' Squadron (Major Loraine-Smith) with one company from each 6/7 Rajputs and the Borders were ordered to move in a left hook and establish a road block at M. S. 6. Meanwhile 'C' Squadron was to move forward along the road and, having cleared the block, was ordered to clear the airfields on both sides of the road. 'A' Squadron carried out its task with success although the ground was difficult, and was established on the objective by 1345 hours. 'C' Squadron, deploying on both sides of the road, attacked the road block. The bridge and nullah on each side of it were extensively mined, and the North end of the nullah was defended by a sniper screen in the bushes. The snipers fled back into a nullah and it was difficult to engage them on account of the steep banks. Eventually crossings were found on both sides of the bridge, and one tank (Capt Riazul Karim Khan) was driven down the bank of a nullah towards the bridge 'drumming up' the snipers in the nullah. The remainder of the squadron then crossed and moved up the road deploying on both sides. A few enemy positions were engaged and destroyed, but the enemy offered no real resistance, retreating rapidly and being caught in the open by 'A' Squadron. Several LMGs, two battalion guns, and two 75 mm A. A. guns were captured. This force then harboured at M. S. 6 for the night, a lane having been cleared through the mined crossing by 36 Field Squadron I. E. B1 Echelon did not move up during the evening, as the minimum of soft vehicles were required forward during the attack on *Meiktila.*

Chapter II.

On the 28th February, the force moved to a concentration area at M. S. 6 preparatory to the attack on *Meiktila*. The plan was that 255 Indian Tank Brigade should carry out a flanking movement to the North East and East of the town prior to attacking it from the East. Meanwhile 48 Indian Infantry Brigade was to attack from the North and 63 Indian Infantry Brigade from the West.

Two recce groups were given the task of recceing the routes into *Meiktila* from the North East and East respectively. 'A' Recce Group reached its objective at M. S. 339½ on the *Mandalay* road without opposition. Its task was to demonstrate in the *Kyigon* area. 'B' Recce Group met no opposition until it had crossed the airfield and . as South of the railway line due South of *Khanda*. This group was closely followed by Probyn's Horse and the 6/7 Rajputs, who in turn were followed by Tac Brigade H. Q. and the remainder of the 'A' echelon vehicles. By this time 'B' Recce Group was hotly engaged South of the railway line. The opposition came from thick scrub covering the canal and level crossing area. The tanks in this group moved to Point 860, and its infantry established a firm base covering the left flank It was hoped that these tanks would be able to support an attack by Probyn's Horse and the 6/7 Rajputs from the East, with an F. U. P. just East of *Khanda* and an axis astride the railway, the limit of the objective being the canal joining the two lakes. On arrival at the F. U. P. it was found that the area South of the railway was under heavy fire which later died down as the result of good shooting by the Recce Group's tanks.

The attack by Probyn's Horse was to be carried out on a two squadron front:—'B' Squadron (Major Stewart) on the right, and 'C' Squadron (Major Arkinstall) on the left. 'B' Squadron deployed without incident, but 'C' Squadron moving across the railway to deploy, came under heavy M. G. fire from the nullah. The infantry sustained casualties and were unable to advance. 'C' Squadron then sent one troop to the nullah and a second troop (Capt Anderson) about 300 yards down road. He halted to see how the troops in the nullah were progressing. He reported that he thought that very few enemy were there, but petrol barrels were being ignited round his tank, which became endangered, and he opened up to get a better look, and was shot through the head. The squadron 2nd in command (Capt. O' Beirne-Ryan) went up to assist in locating the enemy position. Meanwhile the left troop heavily engaged the bunkers in the nullah and destroyed several of them. The infantry again started to advance supported by No 4 Troop (Lieut. Elder), but they received casualties and withdrew. By this time it was considered too late to continue the attack over such difficult ground containing thick trees houses and nullahs which were obviously very strongly held. The infantry, therefore, were left North of the railway while two squadrons of tanks moved to a point just North of *Khanda*, one squadron across the main *Mandalay* road, one squadron to the nullah running North and South a thousand yards West of *Khanda*. The third squadron went again up the nullah South of the railway and had another shoot on to enemy positions. After this the Regiment returned to harbour West of the airfield. Next morning patrols reported that the enemy had vacated the nullah position.

On the 1st March, 48 Indian Infantry Brigade had moved forward to the North and North East of *Meiktila*. 63 Indian Infantry Brigade had moved West of the Lakes and were threatening the Western half of the town from North of *Khanna* village. 'A' Squadron Probyn's Horse (Major Loraine-Smith) moved out and joined 63 Indian Infantry Brigade. 'B' Squadron moved up to Point 860 in order to relieve a squadron of the Royal Deccan Horse, with orders to find a crossing over the escape channel near the rifle range. Of the two bridges over the escape channel on the main road just South of Point 860, one was prepared for demolition and the other destroyed. 'B' Squadron (Major Stewart) spent some time in mopping up bunkers in the Point 860 position. A sapper recce detachment was brought up, and, with fire support from 'B' Squadron and the squadron of the Royal Deccan Horse which had been unable to cross the escape channel, a company of the 6/7 Rajputs was sent across. The intact bridge was them cleared and two troops (Capt Babar and Lieut Grover) with two platoons of the 6/7 Rajputs crossed the bridge, which was only Class 12, and halted on the high ground just South of it. Some enemy remained in the nullah East of the destroyed bridge and were not liquidated for two or three days. On arrival at the high ground South of the bridge, the two troops spotted some enemy and a gun position which they engaged on Point 799.

At 0630 hours on the same day, No 1 Troop of 'A' Squadron (Lieut Lane) was ordered to move round the North of *Meiktila* to join 63 Indian Infantry Brigade. This was a long journey, as *Meiktila* was not then clear, and the only route lay round the North lake. This troop found the brigade North of *Kanna*, and at about 1215 hours attacked *Kanna* from the North. The troop was supported by two companies of the 1/10 Gurkha Rifles. The tanks lead the infantry and the village was cleared first up to the railway line, and then up to the main road. From the main road Japs were seen with-drawing towards the lake. O. C. 1/10 Gurkha Rifles then ordered one tank to protect each flank of the main road and one tank to move on the main road towards the lake. The tanks were therefore operating independently and control could not be exercised by the troop leader. The left hand company was held up by sniping from a gaol, and the troop leader with one company went off to clear this area. Meanwhile the right tank (Dfr Surat Singh), directed by the infantry, had been bunker busting in a hospital compound which contained many bunkers. As a result of this action 120 Japs were killed. In the meantime the remainder of 'A' Squadron had been ordered to report to 63 Indian Infantry Brigade, and on arrival were given the task of supporting an attack by one company of by the 7/10 Baluch Regiment on *Magyigon* village from M. S. 321 on the railway. This was carried out with No 4 Troop (Lieut Bahadur Singh) on the right and No 2 Troop (Ris Siri Singh) on the left, the squadron H. Q. tanks being in the centre. The tanks moved under an artillery concentration and gave fire support to the infantry entering the village. 30 Japs sheltering under a railway culvert were engege by Capt Nicolson (squadron 2nd in command). On the capture of the objective the whole squadron harboured South of *Kanna* between the railway and the main road.

At 0650 hours on the 2nd March 'A' Squadron, in support of one company 1/10 Gurkha Rifles, was given the task of clearing the goal area from the road running South of *Kanna* to the causeway. The attack was carried out with three troops up, and the infantry preceeded the tanks until held up. A mortar concentration was put on to the gaol whose wall was breached by the tanks. The infantry were then sent inside but the gaol was found to be empty. On reaching the causeway, both Capt Nicolson and Lieut Bahadur Singh's tanks received direct hits from a 75 mm gun firing from the far end of the causeway. Perfect communications were maintained throughout between the squadron commander's 38 wireless set and the company commander's 48 set.

It was decided next to clear the *Kyaukpu* area, which was strongly held and consisted of dense thorn thickets and well built stone buildings containing vehicle pits, fox holes and air raid shelters. The F. U. P. was in the hospital compound, which now contained blazing ruins and was strewn with corpses, the result of the previous day's battle The axis given was at an angle to the main road, but visibility on the left was limited to 50 yards by thorn thickets. On reaching the F. U. P. the infantry were pinned by LMG fire. 'A' squadron was in support of 'A' company of the 7/10 Baluch. The Squadron was formed up 'three up' with squadron H Q. behind the left and centre troops. There was some delay through waiting for an airstrike which did not take place. A barrage was then put down by the divisional artillery but appeared to do but little damage. As soon as the artillery fire died down the tanks advanced, the left troop finding itself unsupported by infantry in thick scrub honeycombed by bunkers. This belt of scrub was a hundred yards deep and contained a few houses. As the tanks entered the scrub they came under a hail of small arms fire, and Jem Fateh Singh's tank was assaulted by tank hunters, who placed a charge on the track disabling the tank, and hurled petrol bombs onto it. These Japs were killed by co-ax fire from another tank (Jemadar Sarup Singh). No 4 Troop leader's tank (Lieut Bahadur Singh) was similarly attacked, but no damage was sustained. The tanks continued to blast methodically each bunker which they could see. The left troop and half squadron H. Q. advanced slowly through the scrub until it came out on the far side where it killed many Japs who were breaking and running into a nullah. They were killed in the open. No 4 Troop moved half a mile beyond the objective, engaging Japs in the open and in the hedgerows as they withdrew. This troop and half squadron H. Q. were then ordered to join the right half squadron, who had been progressing steadily but slowly, destroying bunkers the whole way. The infantry here were also unable to keep up with the tanks owing to snipers and LMG fire. Several snipers were knocked out of trees by the tanks and one was brought down from a platform 40 feet high. Resistance finally centred in one large red house which contained a deep air raid shelter. Here too the forceful action of the tanks broke the enemy's morale and they fled from the position, being heavily engaged by No 3 Troop (Ris Bhag Singh) on the *Kyaukpadaung* road as they withdrew into the open. Numbers of enemy withdrawing down the lakeside were engaged by No 2 Troop of 'C' Squadron (Lieut Elder) from Point 799 on the opposite side of the lake. This troop had fine 'Ahmednagar' shoots at 2,000 yards range. 'A' Squadron claimed to have killed at least 300 Japs. This concludes the action with 63 Indian Infantry

Brigade, and the squadron rejoined the Regiment the following morning.

'C' Squadron, meanwhile, had been operating with the 6/7 Rajputs in the area of Point 860 and had cleared up isolated pockets of resistance. Throughout the day one troop had been established on Point 799 over looking the South and West sides of the lake.

On the evening of the 2nd March, O. C. Probyn's Horse was ordered to clear up enemy reported to be on the aerodrome near M. S. 343 on the *Mandalay* road. Information was very scanty, but the enemy position was indicated as being about 500 yards South of M. S. 344. The force was to consist of one company of Sikh Light Infantry, one company 6/7 Rajputs, 'C' Squadron Probyn's Horse, two troops of armoured cars. In support was 59/18 Battery R.A., a V. C. P. and a battery of field artillery. An armoured vehicle of the 16th Light Cavalry had been knocked out at M. S. 344 by what was believed to be an anti tank gun. The 'meet' was at the road block at M. S. 342 at 0700 hours on the 3rd March. The commanders plan was to do a left hook onto the road junction at M. S. 344½, and then to sweep back South of the road in order to take the enemy in the rear.

The first bound selected was at the dispersal area 'A' (see sketch map). One troop of 'C' Squadron (Capt. Riazul Karim Khan) was sent to secure this bound and, having reported it clear, to move to *Myaunggan* village and to report that clear also. Both companies of infantry were carried on the tanks, with the exception of one platoon which had to follow as best it could. On *Myaunggan* being reported clear the Sikh Light Infantry moved from 'A' to 'B' on their feet and formed up ready to move down onto the road junction if required. Meanwhile No. 4 Troop of 'C' Squadron (Lieut. Elder) was directed onto the road junction at 'C', and No. 2 Troop (Lieut. Edwards) onto the end of the airstrip at 'D'. No. 4 Troop was engaged by a 37 and a 75 gun as it moved forward, and the infantry of both troops dispersed and took cover. The tanks engaged the two guns which were located South of the road; both were silenced. It was considered that, as yet, there was insufficient information to use the air or the artillery. No 1 Troop (Capt. Riazul Karim Khan) was therefore directed onto the road junction at 'C' from the area 'B'. This troop was fired on by two 75 mm guns located on the road at 75 yards range. A shell struck the armour on the top of the off side petrol tank of Dfr. Mohd. Sarwar's tank, dented the armour, and set fire to the clothing bin. It is believed that the shock burst the petrol tank because the engine compartment caught fire and the extinguishers failed to put it out because the fumes were reignited from the clothing bin. Meanwhile repeated H. E. bursts were hitting the tank. The guns were engaged, and covering fire was given to the crew enabling them to evacuate without casualties. The two enemy gun crews abandoned their guns. The extent of the enemy position could now be visulised, so No 1 Troop was withdrawn and the artillery had an effective shoot just ahead of the tanks. The 6/7 Rajputs were sent forward to the line of the road, and found the position to have been destroyed. They were ordered to consolidate the road junction, and the Sikh company was pushed across the road into *Shande* village as quickly as possible. This was found to be clear except for enemy equipment. With very little delay the squadron reformed and continued the sweep back along the South side of the road with no opposition. Meanwhile the sapper detachment started to clear the road block. The armoured cars were sent through and almost immediately reported another road block at M. S. 346½. The command and V. C. P. tanks went up the road to join the armoured cars and enemy could be seen moving in the bushes. They were engaged immediately and an airstrike was put in within a few minutes, the target being indicated by W. P. smoke. By this time the infantry had cleared the aerodrome. It was too late to continue the action, and the squadron returned to camp with its infantry, a new road block being established by the 6/15 Punjab Regiment at M. S. 344½. The enemy did not stay to fight and only ten dead were picked up. However three 75 mm and one 37 mm anti tank gun were captured and brought in.

At 0945 hours on the 4th March, No. 4 Troop (Lieut. Carey) 'B' Squadron received orders to proceed to the track junction 800 yards due South of *Okshitkon* village and there to report to the 12th Frontier Force Rifles. After explaining the situation to Lieut. Carey, the officer commanding ordered him to attack *Okshitkon* with one company of infantry from a South Easterly direction working North West. Lieut. Carey moved his troop up to the rear of the company which was already in position. While he talked to the company commander enemy sniping was spasmodic. The attack went in, but 20 yards inside the village the infantry were pinned to the ground. One of the escort to Lieut. Carey's tank was killed and another wounded. His other tanks were also being

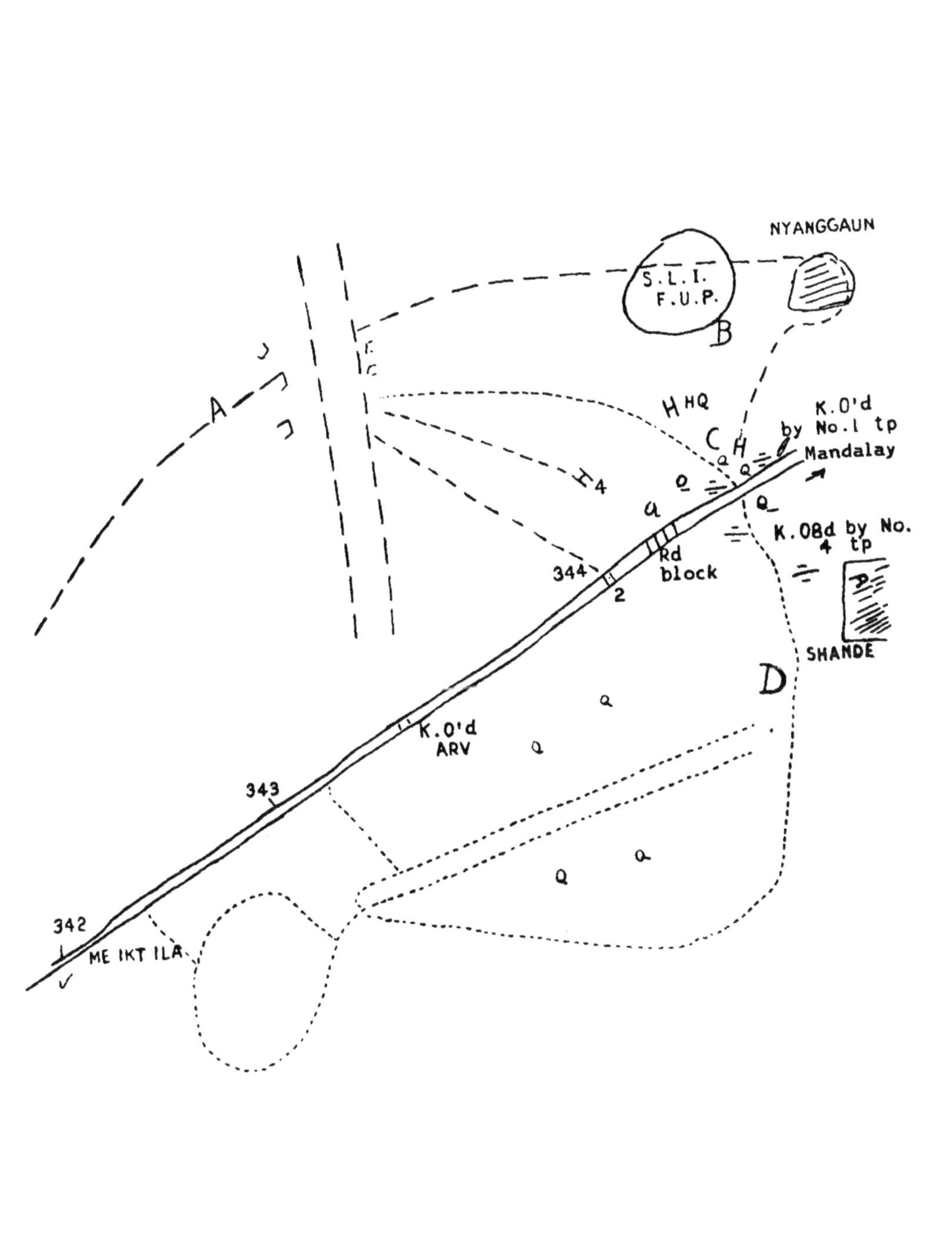

NYANGGAUN
S.L.I.
F.U.P.
B
A
H HQ
C
H
K.O'd
by No.1 tp
Mandalay
I 4
K.O8d by No.
4 tp
344
2
Rd
block
SHANDE
D
K.O'd
ARV
343
342
MEIKTILA

A D-8 (TANK RECOVERY TRACTOR)

shot at. The troop closed to within ten yards of the edge of the enemy's position and engaged several bankers. Lieut. Carey's tank, covered by the other two tanks, entered the position and engaged further targets, blowing up a dump in the centre of the village. Lieut. Carey them withdrew himself leaving two tanks in position. He attempted to smoke the position but without success. He then withdrew further and arranged with the company commander to attack with a platoon from further East supported by his own tank. Again the attack failed and the infantry were pinned to the ground. Lieut. Carey then ordered his other two tanks to try and flush by fire the enemy opposite him. This was fairly successful. The tank and infantry approached the South East corner of the village. Suddenly the lap gunner reported a 70 mm gun between two foxholes at 80 yards range. This was blown over and its crew of ten were killed. Further foxholes in the vicinity were engaged. Thirty minutes later, Lieut. Carey engaged a party of Japs who were retreating Eastwards across his front at regular intervals. When they ceased he withdrew his troop towards the company commander who informed him that the operation was to be abandoned. Estimated damage caused to the enemy was at follows:—

1 70 mm gun destroyed.

2 LMGs destroyed (later confirmed to be 3).

40 killed.

From information received it appeared that the enemy were building up a force astride the *Mandalay* road. It was decided, therefore, to send a force to clear up enemy encountered in the *Okshitkon* and *Nyaungaing* areas. A second force, under command of Lieut. Col. Smeeton, was ordered to clear the area in the vicinity of *Ywadan* on the *Mandalay* road. The operation was highly successful. A description of it is as follows:—

It was known that armoured car patrols had been fired at by mortars between *Kyunbobin* and *Nagyigan*, and by 4 inch mortars in the *Ywadan* area. It was also known that a road block at M. S. 346 was still held. Therefore O. C. Probyn's Horse with 'A' Squadron (Major Loraine-Smith), one company 6/15 Punjab Regiment, two armoured car patrols, one V. C. P., one battery of mountain artillery and the divisional artillery in support, was ordered to clear this area and send patrols North East and East up the *Wundwin* and *Hanza* roads. The force commander's plan was to destroy the enemy by moving in a wide left hand sweep round the suspected area.

The 'meet' was at 0730 hours on the 5th March. Infantry were taken onto the tanks, the mountain guns put into action, and infantry lorries dropped at our road block at M. S. 344½. The mountain gunner was now in position and protected, vehicles not required had been shed, and the force started off across country. The villages of *Magyigan* and *Kyunbobin* were first reported clear, a troop going to each and putting through its infantry platoon. *Tanaggan* was searched next and was cleared of a few enemy. As the leading troop (Ris Siri Singh) moved to the pagoda near the village, three 75 mm shells were fired at it, giving a perfect resection to the road junction and betraying the main enemy position. No 3 Troop (Ris Bhag Singh) pushed due East onto the main road in order to cut off any withdrawal. No 1 Troop (Lieut Lane) was sent on towards the cross roads in order to draw enemy fire, while No 4 Troop (Lieut. Bahadur Singh), Squadron H. Q. and the command tank watched from the pagoda area. No 1 Troop was soon very hotly engaged, three gun positions being spotted from the pagoda and engaged by No. 4 Troop and the command tank. No. 1 Troop was then halted and a most affective air strike went in on the target indicated by the command tank. Indication was done by H.E. Immediately after the airstrike, No. 1 and No. 4 Troops followed by infantry were ordered to capture the enemy position in the road junction area up to the line of the road. Some snipers were encountered and killed in bushes in front of the position, but the infantry were quickly pinned by M. G. fire from prepared but inadequate positions in a high bund built round some pagodas. No 4 Troop went close up to the bund and No 1 Troop moved down the side of the bund to the road hammering the position. Lieut. Lane crossed the road and was hit several times at a hundred yards range by two 75 mm guns located in the bushes South of the road. His engine was set on fire and his tank received a complete penetration in the left sponson in front of the ammunition bin, which set the dashboard wiring on fire. Both fires were extinguished, but the engine had stopped and the wireless was out of order. The tank remained in action, using manual control, until all opposition ceased, and the troop leader had destroyed two guns from his own tank. The second tank came up to give support, and the crew of the first tank evacuated without casualties. The squadron 2nd in command (Capt. Nicolson) with a squadron H.Q. tank had placed himself on the left of No. 4 Troop, and could see into the rear of the position. The enemy then started to try and vacate it by crawling along the ground or running from bush to bush. Both Capt Nicolson and No 2 Troop, who were on the road, carried out great slaughter, bodies in some cases were piled on the top of each other. The infantry, who had sustained some casualties, went in but came out again as there were still enemy in a bunker behind the bund. This bunker, having been located, was destroyed by the tanks and the enemy behind the bund were 'hotted up' with grenade dischargers. The infantry went in a second time and found the position to be unoccupied.

Meanwhile a patrol of the 11th Cavalry (P. A. V. O.) had gone up the *Mandalay* road and continuously engaged small parties of enemy, destroying a tankette and capturing a 105 mm gun whose crew were asleep. This patrol showed the most dashing cavalry spirit. A second armoured car patrol reported enemy at the road block at M. S. 346. These were dispursed by the patrol, supported by the mountain guns, observation being given by the patrol to the battery commander with the control tank. This patrol was then sent down the *Hanza* road where it soon came into contact with fleeing enemy. Sappers cleared the roadblock and a D 8 towed away the damaged tank. The whole force then returned to harbour after a very enjoyable day.

Our casualties:	1 killed. 11 wounded. 1 tank out of action.
Enemy casualties:	180 killed (incl 3 officers) 4 75 mm guns 1 bn gun. 1 105 mm gun.

On the 6th March, a force known as '*Ralphcol*' composed of the following units was sent to carry out a recce in force in the area *Ywadan - Hanza - Thedaw - Wundwin* and was ordered to destroy all enemy encountered.

Comd	Col R. Younger (2-in-C 255 Ind Tk Bde)
Tps	Sqn 16 Cav less one tp. Probyn's Horse less one sqn.

21 Mtn Regt, less one bty.
1/7 G. R.
Coy 6/15 Punjab Regt.
Tp 36 Fd Sqn.
Rec and Med dets.

The column met very slight opposition when advancing along the main road, and was able to establish its first firm base in the area of the cross roads at *Ywadan*. Patrols were sent forward to *Shawbin* and *Wuntha*. 'C' Squadron (Major Arkinstall), with one company 1/7 Gurkhas, advanced to search the villages North West of *Ywadan*, and finally to *Shawbin* which was found to be empty. A large dump of medical stores was found under a pagoda. One troop of armoured cars had been sent to *Hanza* and reported that they had heard light automatic fire from the area of the *Hanza* level crossing. 'B' Squadron (Major Stewart) was therefore sent to clear up the situation. The fire was discovered to be coming from some railway carriages These were destroyed and set on fire by No 3 Troop (Lt Grover). The villages of *Hanza* and *Wuntha* were found to be empty, and the infantry company was established in *Hanza* for the night. 'B' Squadron returned to the force harbour at *Ywadan* at 1830 hrs. A jeep recce from the Scout Troop under Lt Quinn had been sent to follow the armoured cars as far as *Hanza*, and then continued due East to observe the road as far as 8 miles from *Hanza* in order to see whether it was being used by the Japs. He obtained useful information and shot up a party of enemy killing three of them.

On the 7th March '*Ralphcol*' was split into two columns. The first '*Fredcol*' (Major Kennedy), consisting of 'C' Squadron Probyn's Horse one company 1/7 Gurkhas and one mountain battey, received orders to move East of *Hanza*, thence North up the railway to clear the village of *Thedaw*. The company of 1/7 Gurkhas was to form a patrol base in that village and send fighting patrols to *Wundwin*. This was carried out without event except for a short brush with a Jap party during a 'top up'. One Jap officer and one other rank were killed by Capt Riazul Karim Khan. *Thedaw* was found to be deserted, but one tankette, three armoured tractors and one truck were destroyed.

The second column '*Chapcol*', under command of Lieut-Col Chaplain R. A. and consisting of one troop 16 Cavalry, one troop of 'B' Squadron Probyn's Horse (Capt Babar), one company 1/7 Gurkhas and one mountain battery, was ordered to capture a large enemy dump at *Shawbin*, then to move as far as *Wundwin* clearing any enemy encountered. The column arrived at *Shawbin* without opposition. The locality of the dump was discovered by questioning a local inhabitant. It was found to be in a large hut and contained medical stores and equipment. At about 1030 hrs, as these were being loaded into trucks, the armoured cars sent a message that about ten Japs were moving down the East side of the road which was covered by thick scrub and bushes. Capt Babar went at once to meet them. As soon as they came into view he opened fire; the enemy withdrew at once into a nullah. At 1300 hrs the armoured cars sent another message that about fifty Japs were coming down the same road. Capt Babar, with a platoon of infantry on his tanks, had advanced about a mile before the enemy opened fire on him. He immediately took up a position on the East side of the road and engaged the enemy at a range of sixty yards. Visibility was bad and there was a deep nullah to his front. He engaged the enemy until 1630 hrs and killed a number. The nullah was an anti tank obstacle and the only way of crossing was by a bridge. He was ordered to cross it but refused to do so until the sappers had stated that it was strong enough. After much argument this was done and the tanks and infantry crossed. Capt Babar saw more enemy whom he engaged. The infantry then went forward and mopped up. Thirty dead were definitely counted on the ground, but it was estimated that at least forty had been killed. The troop returned to harbour at 1800 hrs.

Lieut Bailey repeated and completed the recce carried out by Lieut Quinn on the previous day. He obtained useful information and mined the road.

On the night 7/8 March, patrols reported that *Wundwin* was only lightly held. At 0900 hrs the following morning '*Ralphcol*' withdrew to *Meiktila* having achieved its object, and having obtained a considerable amount of valuable information. It is estimated that 70 enemy were killed and several vehicles destroyed including one tankette and three gun tractors.

It was now clear that the rapid advance to and capture of the town of *Meiktila* had completely disorganized the enemy's defences, communications and L of C for some considerable distance round the town. For at least a week there appeared to be no higher command controlling the small parties of enemy in the area, who appeared to wander from village to village with no particular object in view. Several half hearted attacks were put in onto the defences around *Meiktila*, but as these were unsupported they were easily repulsed. On about the 10th March the enemy began to close in round the town, and his troops were obviously being organized into a force. One Jap division occupied positions North of *Meiktila* and infiltrated into the untankable area North East of the North Lake. At the same time a Jap regiment with artillery was moving up from the South and arrived in the *Yindaw* area on about the 11th March. On the 15th March the enemy began to evolve two intentions: firstly to deny to us the use of the *Meiktila* Airfield, and secondly to contain the *Meiktila* forces within the *Meiktila* area. The first of these tasks the enemy succeeded in carrying out until the 30th March, but his force was too weak and he was unable to put more than a thin screen supported by a few guns round *Meiktila*.

Information was required as to whether of not the *Mahlaing* road was clear in order that 17 Indian Division's B 2 and B 3 echelons might move up to *Meiktila*. Therefore, on the 7th March, one squadron of the Royal Deccan Horse with two companies of the 4/4 Bombay Grenadiers, under command of Lieut Col. Tighe, was ordered to reconnoitre up to the *Thinbon Chaung* at M.S. 11½. Slight opposition was met at the villages of *Leindaw* and *Kyaungyagon* and at the village North of M.S. 8½. On arrival in the area of *Yegyo* the force was engaged by small arms and gun fire. One 37 mm gun was reported to be in the North corner of *Thangongya* village beside a hay stack which was engaged and set on fire with the most surprising results. It was concluded that an ammunition dump must have also been hidden in the hay stack. By this time it was getting late and so Lieut Col Tighe, finding that the enemy had returned to positions near M. S. 6, decided to withdraw. The column was shot at as it moved back along the main road.

On the 8th March the following force was sent out to clear up the *Yegyo* area and exploit as far North as M.S. 13½:—

Comd. Brigadier Hedley.

Tps. 'A' Sqn Probyn's Horse.
48 Ind Inf Bde less one bn and one coy.
Bty medium arty.
Two btys fd arty.

Major Loraine-Smith was put in command of the advance guard, which consisted of his squadron and a company of the

A TANK TRANSPORTER TOWING
A LEE TANK.

West Yorks. The 'meet' was at 0645 hours at the monastery North of *Meiktila* on the *Mahlaing* road. The vanguard consisted of eleven tanks and one platoon of infantary which was mounted on the tanks of No. 3 Troop (Ris Bhag Singh). The remainder of the company was lorry borne No opposition was met until the vanguard reached M. S. 6, when No 4 Troop (Lieut Bahadur Singh), which was leading, was fired at from the villages to the North. This troop cleared *Kyaungyagon* with the assistance of No 3 Troop and the platoon of infantry. No 4 Troop was sent round *Leindaw*, at the East end of which a party of 30 Japs was seen approaching from the East. These were engaged at 500 yards range and fled in the direction from which they had come pursued by No 4 Troop who claim to have killed 15. The brigade commander left one company of infantry to clear this area.

The advance continued in the formation 'two up', No 4 Troop right and No 2 Troop left. A party of 20 Japs was engaged and killed at the road and nullah crossing at M. S. 8½. Ten were killed and the remainder dispursed in the scrub. Movement was hindered by the presence of aerial bombs which had not been removed since the action on this road during the advance on *Meiktila*. No 4 Troop, meanwhile, crossed the nullah and engaged enemy in the area of *Ledaingzin*, killing about 20. No 2 Troop was next directed on to the road junction North West of *Yegyo*, while No 4 Troop, on its way to *Thangongyi*, saw a party of 150 Japs in the open and succeeded in killing 40. Near this village a burnt out Jap tank was found which had obviously been concealed in a haystack and destroyed previously by the Royal Deccan Horse. No 2 Troop was sent next to to recce the village of *Thabyeban* and the nearby causeway, and No 3 Troop with infantry was put through *Yegyo*. At M. S. 10½, No 3 Troop was engaged by gunfire which appeared to come from the area of the nullah bend 1500 yards North East of the causeway. No 3 Troop was consequently ordered to change its axis of advance parallel to the nullah and to advance North East. Meanwhile No 2 Troop crossed the nullah and came upon a 105 mm gun, heavily camouflaged, 200 yards from the causeway; with it was an ammunition lorry. These were engaged and destroyed by Jem Phula Singh, and the troop continued North across the nullah engaging five trenches near the gun position which contained 100 enemy of whom 30 were killed. Meanwhile enemy had been spotted from *Yegyo* in the area of an elbow bend in a nullah 500 yards East of the gun position. These were engaged by Squadron H. Q. and No 3 Troop, the latter at a range of 300 yards. The enemy fled across the open in a Northerly direction towards *Ywathit.* No 4 Troop and Squadron H. Q. then moved up to the nullah elbow, No 4 Troop being directed West along the bank of the nullah, while No 3 Troop kept observation on *Ywathit.* No 4 Troop spotted and engaged a 105 mm gun and tractor with a crew of ten who were moving up the West bank of the nullah in the direction of *Ywathit.* The gun and its crew were destroyed. No 4 Troop crossed the nullah and located a 75 mm gun with tractor and an ammunition lorry 50 yards West of the first gun position. This together with a nearby ammunition dump was destroyed. By this time it was 1515 hours, and the progress possible across country was very slow. The squadron moved back to *Yegyo*, less No 2 Troop and one platoon of infantry, who recced as far as M. S. 13½ without meeting further opposition. Meanwhile a party of Japs had returned to the area of M. S. 6 and again sniped the road. They were dealt with by a company of infantry. The column then returned to *Meiktila*, employing one troop to support a lay-back at M. S. 6 and another troop as rearguard. Harbour was reached by 1830 hours.

A force, under command of Lieut Col Smeeton, was ordered to move towards *Mahlaing* with the object of escorting 17 Division's B2 and B3 Echelons to *Meiktila.* This move was to take place in conjunction with an operation by 33 Indian Infantry Brigade South of of *Taungtha.* Only slight opposition was expected, although little was known of the enemy situation along the road. The 'meet' took place at 0645 hours on the 10th March at the road Junction at M. S. 2. The force moved off in the following order:—

Advance Guard Mounted Troops.

Comd	Lt Col Chaudhuri.
Tps	Three tps 16 Cav. One tp Probyn's ('A' Sqn).

Advance guard.

Comd	Maj Stewart.
Tps	One sqn Probyn's ('B' Sqn with one coy 4/12 F. F. R. less one pl on the tks) F. O. O.

Main body

Tps	Column H.Q.
	4/12 F. F. R. less three coys.
	'A' Ech tpt.
	Fd Amb.
	One pl 9/13 F. F. R. (MMG.)
	2 Fd Bty I. A.
	One pl 4/12 F. F. R.

Rear Guard.

Comd	Major Amrik Singh.
Tps	One sec 21 Mtn Bty.
	Onecoy 4/12 F. F. R.

The first enemy reported were in the scrub at M. S. 5 on both sides of the road, and the troop of 'A' Squadron with the armoured cars engaged and destroyed an anit tank gun sited behind a small bund. The advance guard was then deployed and moved up on both sides of the road with two troops up and one in reserve actually on the road. The infantry soon came under small arms and 75 mm gunfire from an unlocated gun position. The advance was very slow communication between tanks and infantry being bad. The tanks on the right of the road soon found themselves well ahead of the infantry. They engaged many enemy in the bushes and in hastily prepared positions on a bund. The main enemy position appeared be on the high ground at m. s. 5½, and No 4 Troop (Lt. Carey), No 3 Troop (Lt. Grover) and Squadron H. Q., less Capt. Udham Singh, of 'B' Squadron, together with a Bombay Grenadier e scort, were then sent forward with orders to move North of the road, cross it at m. s. 6, and pass behind the enemy position coming in on his right flank. Meanwhile No. 1 Troop (Capt Babar) and Capt Udham Singh remained on the left of the road, engaging the enemy to their immediate front. No 2 Troop of 'A' Squadron (Ris. Siri Singh) and No 2 Troop 'B' Squadron (Ris. Shamsher Singh), under command of Capt. Nicolson, moved out onto the right flank to try and engage 105 mm guns believed to be in that area. This encircling movement was successfully carried out, and the enemy vacated their positions as the two troops came up on the flank. The enemy ran North across the road towards the village of *Leindaw*, and were caught in the cross fire of the troops on their front and rear. On reaching the top of the hill Nos 3 and 4 Troops of 'B' Squadron found many enemy in deep trenches, a number of which had been dug by our troops when they harboured there during the advance on *Meiktila.* 'B' Squadron commander reported that he would need infantry to clear the position. The leading company had sustained a number of casualties due to sniping and M. G. fire, so it was decided to employ another company to carry out this operation. No 4 Troop was called off the hill while No 3 Troop and Squadron H. Q. remained in observation. No 4 Troop 'married up' with the second company and moved forward to mop up. They reached the top of the hill without opposition and started to clear it when an enemy 105 mm gun opened up. The infantry moved off the top and, as it was feared that the tanks would sustain casualties, they too were ordered to pull back from the crest. The tanks had just completed 'topping up' and, before the petrol lorry had moved off, a Jap jumped out of a slit trench and threw a grenade at it, wounding Jem Mehar Singh in the leg. It was now obvious that the enemy gun positions would have to be cleared up before the hill could be occupied, and for this task a third company was 'married up' to 'A' Squadron and sent round *Leindaw* in a wide sweep. Meanwhile 'B' Squadron commander, finding that the petrol lorry and been left behind, because it it could not be started owing to damage caused by the grenade, went back with the echelon officer (Lieut Turner) to recover it. As Swr Gurbachan Singh was getting into the driver's seat he was sniped from a trench and killed. At the same time Swr Karnail Singh was wounded. No 4 Troop was called up to give protection while the lorry was being recovered. When organizing the tow of the petrol lorry, Lieut Turner saw two Japs in a trench and fired at them with his revolver. He also threw a grenade into the trench which one of the Japs promptly threw back at him. He then threw in a second grenade which killed the Japs. The lorry was towed away. Enemy shellfire onto the hill continued, and as it was getting late Lieut Col Mc Leod of the 4/12 F. F. R. was ordered to pull back his leading company and organize a harbour for the night.

Meanwile the 'A' Squadron tanks with their supporting infantry had been making good progress and destroyed a 37 mm and a 75 mm gun. Under the cover of an airstrike they succeeded in getting right up to *Leindaw* village, and Japs were found to have fled northwards, leaving telephone wires in position. 'B' Squadron and the second company of infantry, having completed their mopping up, had moved back to harbour and the right company was withdrawn. It was estimated that at least 150 Japs had been killed, and the ground was littered with corpses. It was not anticipated that the position would be reoccupied. The force spent a quiet night in harbour until 0400 hrs, when a Jap patrol consisting of about 15 men walked into the perimeter. Five were killed a few yards outside the defences.

At 0645 hrs the next morning, two troops were ordered to move forward onto the high ground and, if this area was found to be clear, to recconoitre *Leindaw* and *Kyaungyagon* prior to the continuation of the advance to *Mahlaing*. A fighting patrol, which had been in position on the hill during the whole night, returned and reported that up to 0400 hrs no enemy had been seen. However just before the two troops moved off enemy were seen to be digging on the top of the hill. This was thought to be a new force which had moved across from the *Pindale* road area. Shortly after they were seen digging in the bushes 400 yards from the harbour. Two troops were despatched immediately to deal with them and give room for the deployment of the force. The enemy at once withdrew from our front. Ammunition vehicles had now arrived from *Meiktila*, and it was decided that, before continuing the attack onto the hill, which appeared to be a necessary operation, the tanks should replenish ammunition expenditure. When this had been done the infantry 'married up'. The attack was to be supported by artillery and an airstrike, but the force had hardly moved forward before snipers opened up from the bushes, as they had done on the previous day. This time the tanks remained with the infantry who were sustaining shell fire in addition to small arms fire. It was decided to clear up the left flank, where the infantry held up, before the advance continued. Consequently one troop of tanks and a company of infantry were sent round the left flank.

Meanwhile 'A' Squadron's tanks without infantry were sent out on the right to make an attempt to locate the enemy's 75 mm guns. The attack on the left progressed well and as the tanks came up with the leading company it was ordered to advance. They had not moved 100 yards before the 75 mm guns opened up, supported by a 105 mm gun. The infantry were pinned to the ground and, in order to avoid casualties, both tanks and infantry were ordered to pull back away from the shelling which ceased immediately.

The troop on the right had approached *Leindaw* by this time, but as soon as it 'showed its nose' above a ridge South of the nullah it was heavily engaged by a 75 mm gun. An air O. P. was called up, and much time was wasted while this troop advanced to the crest to draw fire and then withdrew. The air O. P. failed completely to locate any gun positions. During this time another force containing 'C' Squadron Probyn's Horse had been sent up the *Pindale* road, but it made no progress, the tanks coming under continuous shell fire.

Although a large number of enemy had been killed on the previous day, it now appeared that he was in far greater strength than had been invisaged. This fact was reported to Brigade H. Q. with the suggestion that the tanks alone should carry out a wide sweep in order to try and ascertain the extent of the enemy's positions, and destroy the guns. This proposal was agreed, and the whole force was ordered to return to *Meiktila* on completion of the sweep. Some delay was caused on the left flank by snipers who were troubling the infantry and had to be killed. An airstrike was put on to the top of the hill followed by an artillery concentration. The armour then moved off in a wide sweep to the left. 'B' Squadron leading, followed by R H. Q. followed by the detachment of 'A' Squadron. A forward wireless link was left with the battery commander. As the enemy position was passed the tanks swung round in a 'two up' formation and came in slightly to the rear. Thereupon the enemy jumped out of their positions and streamed across the road. As they ran they were caught by the artillery directed by the command tank. The infantry still in the previous night's harbour area, stood on their vehicles and cheered the spectacle like a race horse crowd. A gun position was located behind the hill on the South side of the road and was engaged by the tanks. One 75 mm gun was definitely destroyed, and one 75 mm and one battalion gun were probably destroyed. The tanks then swung left up the road to M. S. 6½. They then turned right along the edge of the airfield passing North of *Kyanngyagon* village. A 47 mm anti tank gun was seen on the East side of the village which was engaged and probably destroyed at a range of 800 yards. Before this gun was destroyed it fired very accurately and hit one tank twice on the turret, pinning the mantlet to the turret so that the gun was unable to elevate or depress.

The sweep then continued, and between two villages another 47 mm anti tank gun was spotted and probably destroyed. This gun never returned our fire. As the tanks passed North of *Leindaw*, several were hit by yet another anti tank gun sited in the village. The command tank received a complete penetration through the additional armour protecting the ammunition big. A second R. H. Q. tank was also hit in the same place, but the armour was not penetrated. The anti tank gun was engaged but the result could not be seen owing to dense trees. The rearmost tank commander of 'A' Squadron reported that his track had been shot off, and artillery were called on to give support while the tank was being recovered. A concentration came down almost immediately, the first shot landing with the most surprising accuracy. Thereafter corrections were given by the command tank. The result was that not a single shot was fired from the village while the tank was being towed away. The tanks then moved back to the main road and thence to the harbour area. During the sweep the enemy, apart from the anti tank gunners, appeared to have been completely demoraliszed and ran here and there usually in the direction of villages where they were caught by artillery fire. On arrival back in harbour, it was found that a considerable amount of damage had been done by anti tank fire to suspension assemblies and tracks etc. Four 'lame ducks' eventually waddled home in the dark.

In conclusion, this action was thoroughly enjoyed by both those who took part and by those who watched it. There was insufficient time to press home the attack and ensure the destruction of all the guns. Nevertheless the action certainly had the effect of sending everyone home in the best of spirits.

On the 14th March, it was decided to carry out an armoured sweep round the North of *Meiktila* with the object of destroying enemy when and where met. The force, commanded by Lieut Col Smeeton, consisted of 'A' and 'C' Squadrons Probyn's Horse and the Brigade H. Q. Troop (Lieut Roberton) which was on loan.

Enemy with guns were known to be in force in the area MS. 3 to MS. 5 on the *Pindale* road - *Subyugwet* - MS. 5 *Mahlaing* road. Therefore the axis chosen for the advance was from the harbour area at *Paukchaung* - M.S. 341 on the *Mandalay* road - to the crosstracks West of *Kyibin* - MS. 6 on the *Pindale* road - North of *Leindaw* - M.S. 6 on the *Mahlaing* road.

The Regiment marched at 0930 hours and ran the gauntlet of shelling on the *Mandalay* road. The tanks 'topped up' at MS 341 without incident. At 1145 hours the force advanced with 'C' Squadron leading, followed by R.H.Q. and 'A' Squadron, to *Kyibin* where it changed direction towards MS. 6 on the *Pindale* road. As the leading troop approached the road, small parties of enemy were seen and engaged at long range. 'C' Squadron (Major Arkinstall) was ordered to cross and take up a hull down position west of the road at MS. 5¼. 'A' Squadron (Major Loraine Smith) was ordered to occupy the position vacated by 'C' Squadron. As this Squadron started to move enemy 70 mm and 105 mm guns were located and destroyed by No 2 Troop (Ris. Siri Singh). Meanwhile 'C' Squadron from its hull down position was engaging enemy and gun positions to its front at the South of the causeway. 'A' Squadron moved up in rear facing North and West. The whole area came under intensive shell fire from Point 801, *Leindaw* and Point 827. During this shelling Lieut Col Smeeton, who was forward with 'C' Squadron, was hit by a piece of shrapnel in the nose. Major Loraine-Smith took over command, and Capt Nicolson took over command of 'A' Squadron. The brigadier directed that the sweep should continue. 'A' Squadron was therefore ordered to move due West. The shelling went on and, as the guns could not be located, an airstrike was put down on Point 827 which was a suspected enemy O. P. This was most successful because the shelling ceased and a large amount of smoke and flame completely obscured Point 827 for a considerable time. The advance continued, with 'C' Squadron in the rear, until the leading tanks were within three quarters of a mile from the North East corner of *Leindaw*. These tanks came under shell fire from an unmarked and burnt out village East of *Leindaw* and from a nullah South of it. The enemy guns were engaged by the whole of 'A' Squadron at 600 yards range, resulting in two 75 mm and one 105 guns being definitely destroyed. It appeared that to clear *Leindaw* would take a considerable time, so 'C' Squadron was ordered to pass behind 'A' Squadron to a point overlooking *Kyaungyagon*. Jem Phula Singh's tank had been hit by a 75 mm shell on the engine grill which punctured the oil coolers, and the tank had to be taken on tow. All the Jap guns being silenced and many of the gun crews being killed, it appeared that the area was clear and the advance was continued to M.S. 6 on the *Mahlaing* road. No further opposition was met. An artillery concentration was put down onto the area between *Leindaw* and *Kyaungyagon* as the Regiment passed. From M.S. 6 'C' Squadron was despatched to *Kangaung - Kyaukpon - Kyaukpyugon.* The squadron was fired at by a 75 mm gun at *Kyaukpon* and engaged the gun area but did not observe the effects. 'A' Squadron with three tanks on tow passed East of *Antu* and came under artillery fire from a range of 1200 yards from the area of MS. 5 on the *Mahlaing* road. The guns, estimated at four in number, were all silenced except one. Their crews probably deserted them. Both Squadrons met in *Kyaukpyugon* where an ambulance had been sent to bring in Lieut Col Smeeton. On its was back to *Meiktila* the force was shelled from MS. 3. Lieut Edwards and his driver Swr Shah Hussain were both wounded by shrapnel, but the latter did not report that he had been hit until after he had driven his tank back to harbour. A feature of the day was the small number of enemy seen. Only those surprised were caught in the open, indicating that they had decided to hide and remain hidden when tanks appear. Enemy casualties were estimated as follows:—

30 killed, mostly gun crews.
2 105 mm guns.
3 75 mm guns.
1 47 mm anti tank gun (probable).

On the 15th March, 'B' Squadron (Major Stewart) was ordered to proceed to 99 Indian Infantry Brigade's harbour on the *Meiktila* airstrip. There the squadron was to 'marry up' with two companies 1/3 Gurkha Rifles and establish a firm base at *Kandaingbauk*, thence to demonstrate towards Point 801 with the object of neutralizing enemy guns in that area during the arrival by air of 9 Indian Infantry Brigade. On arrival the brigade commander ordered Major Kennedy, O. C. Force, to clear three villages North of MS. 340, before carrying out his original task; local inhabitants had reported that 600 enemy were there. Actually, these villages were found to be empty and, after considerable delay, the force moved on to *Kandaingbauk*. There two Jap sets of equipment were found which had been recently left behind. Some enemy were pointed out by Major Udham Singh at the South edge of *Shawbyugan*; they were engaged. No 4 Troop (Lieut Carey) moved without infantry towards the high ground at Point 801, and this resulted in enemy shelling opening up onto the North East of *Kandaingbauk*. A probable enemy O. P. was then engaged in the area of the pagoda, North of Point 801. After 'topping up', No 3 Troop (Lieut Grover) moved to the nullah area some 800, yards South of Point 801 and No 1 Troop (Capt Babar) moved round the North of *Shawbyugan* to demonstrate onto Point 801 from the North. Spasmodic enemy shelling continued and some smoke and dust were seen North East of the pagodas which were engaged by the artillery. The shelling ceased immediately. While 'B' Squadron was in this area the enemy did not shell the airstrip. However, as soon as the squadron started to withdraw, a message was received from brigade H.Q. that shelling had again started. No 4 Troop was accordingly sent back to Point 801 with excellent effect. This troop carried out a shoot onto the area of the pagodas, and a dump was seen to blow up in a sheet of flame. The force returned to harbour by 1830 hours, where it was ordered to remain for the night. There was considerable activity around the perimeter during the hours of darkness.

At 2000 hours the same evening, the brigade commander ordered 'B' Squadron to carry out a similar role the following day. Two companies of the 6/15 Punjab Regiment were put under command of Major Kennedy, and 59/18 Battery R.A. together with mountain, field and medium artillery were put in support. The tanks were ordered to move South of the airstrip, thence to the bridge at M. S. 342 in order to avoid drawing enemy shelling on to the airstrip. To save time Major Kennedy ordered one company of infantry with sappers to capture the bridge at first light. The tanks were to move at 0645 hours.

Immediately after 'stand down' infantry patrols and the company of the 6/15 Punjab Regiment moved outside the perimeter; small arms fire opened upon them. After twenty minutes it was apparent that the infantry were pinned to within fifty yards of their own wire. It was decided that the tanks should not be sent out on their mission until the situation had been cleared. No 3 Troop (Lieut Grover) was ordered to remove the enemy from the North end of the airstrip to enable the company to push on to M. S. 342. When the troop reached the gate in the wire, Lieut Grover, being unable to persuade anyone to open it, dismounted from his tank and did so himself. His tank was under considerable fire at the time.

Meanwhile the harbour and airstrip were being shelled from all directions, and it was clear that the original task of the force could not be carried out, as the clearing of the airstrip would require all available troops. Orders for this were issued accordingly, and No 3 Troop cleared the track up to the line of the nullah South of the main road. Here the enemy were dug in both in the main nullah and one the broken scrubby ground at the North of the airstrip, which appeared to be their main position. 'B' Squadron less on troop with one company 6/15 Punjab Regiment moved out of harbour in a 'two up' formation. No 1 Troop (Capt Babar) was on the left, and No 4 Troop (Lieut Carey) was on the right. Slowly the squadron cleared the area towards the main nullah. No 1 and No 4 Troops and Squadron H.Q. had good shoots assisted by the mortars of the mountain regiment for one and a half hours. 59/18 Battery R.A. too had an excellent shoot eighty yards in front of the tanks. 25 dead were counted later. No 3 Troop then moved West down the road towards *Meiktila,* thence South along the edge of the harbour, and finally joined up with the rest of the squadron in the area of the nullah. No 1 Troop leader (Dfr Balbir Singh) located and destroyed a 70 mm anti tank gun. The area was cleared by 1300 hours after which orders were received that one company of infantry should be established in the area of the bridge at M.S. 342. No 1 and No 4 Troops supported this operation without incident. The squadron then rallied and returned to regimental harbour in 255 Indian Tank Brigade area at 1830 hours.

The bag for the day was estimated as follows:—

Enemy killed	200	
Guns destroyed	1	70 mm.
	1	37 mm.
	1	3 inch mortar.
	5	LMGs.
	3	MMGs.

On the 17th March, two squadrons of Probyn's Horse were ordered to support an operation by 63 Indian Infantry Brigade. The task was to destroy all enemy in the areas bounded on the West by the railway *Meiktila—Yegyo* and on the East by the road *Pindale-Meiktila.* The brigade was to sweep as far North as *Sewa,* and the whole operation was to be carried out in conjunction with an attack by elements of 99 Indian Infantry Brigade onto Point 801, supported by one squadron of the Royal Deccan Horse.

The infantry Brigade moved out of harbour at *Meiktila* at 0500 hours and the tanks moved at 0615 hours. The two squadrons were to 'marry up' at *Oknibok* and *Kyaukpyugon.* The advance was then to continue North on two axes: on the right 1/10 Gurkha Rifles, with 'A' Squadron (Major Loraine-Smith) in support, to *Oknebok-Impetlet-Subyugwet*: on the left 7/10 Baluch, with 'C' Squadron (Major Arkinstall) in support, to *Kyaukpyugon-Antu-Kyaungyugon*. Brigade H. Q. moved on the left axis.

'A' Squadron had much difficulty in finding 1/10 Gurkha Rifles who had lost their way in the dark. A Jap tank destroyer party of six men was encountered between *Oknebok* and the lake. Having directed a burst of *LMG* fire onto the Gurkha battalion H.Q. wounding several G.O.Rs., they dived into the lake. All were killed by small arms fire as they were swimming in the water. Villagers reported that *Impetlet* was held by 20 Japs with an *LMG* which was situated in the West corner of the village. The battalion and squadron commanders recced forward to within a thousand yards and decided to attack at 0915 hours, after an airstrike and artillery concentration with two companies up, each supported by a troops of tanks. No 3 Troop (Ris Bhag Singh) on the right, No 1 Troop (Lieut Lane) on the left. The remainder of the squadron was to support from a position 600 yards West of the axis. As soon as the tanks started to manoeuvre, 75 mm supported by 105 mm gunfire opened up from the direction of the pagoda on the East bank of the lake. Gun flashes were spotted and engaged by our artillery. The attack actually commenced at 0930 hours and, in spite of small arms fire and shelling, continued until the edge of the village was reached. It now appeared that this was strongly held and the tanks started firing at enemy bunker positions. The Gurkhas gave excellent target indication and the village was systematically cleared. A centre of resistance was located some 300 yards West of the village, so No 4 Troop (Lieut Bahadur Singh) was sent to clear it with one platoon of infantry. When this had been completed, one company of Gurkhas was sent forward to establish a road block in the area of M. S. 4½ on the *Mahlaing* road, No 1 and No 4 Troops supporting. As they moved onto the high ground North of the village they came under intensive artillery fire from Point 801 and from M.S. 3 on the *Pindale* road. The infantry company commander was killed. The Jap appeared to have no hesitation in shelling his own position between the village and the road. To assist the Gurkhas, No 2 Troop was ordered forward to cut the road at M. S. 5. This troop was engaged by 47 mm and anti tank gun fire. One tank (Ris Siri Singh) was hit on the suspension and was immobilised. The master switch on Jem Jai Singh's tank broke and the tank was temporarily out of action. The Squadron 2nd in command (Capt Nicolson) and the F.O.O. (Capt Allen) were sent round the left to try and locate the guns which they engaged and silenced. The area seemed to be alive with Japs who were behind every bush. Capt Nicolson moved forward towards the road to kill two Japs who were crawling towards him, when his tank was hit by a 47 mm shell, his final drive was penetrated, and his suspension immobilised. He engaged and destroyed the gun. Meanwhile the remainder of the squadron was under shell fire. No 2 Troop supported the company forward to M. S. 4½. Several artillery concentrations were but down, but still the shelling continued. Dfr Amar Singh (No 3 Troop) dismounted from his tank and worked his way well forward to observe. He located two gun areas which were engaged and the guns silenced. The crew of one gun was seen to run away; in the other the gun tender and tractor were destroyed. It was decided next to clear the area as far as *Subyugwet*. No 1 Troop moved forward and engaged enemy in the village, which was set on fire, and killed about 35. No 4 Troop advanced onto some low ground destroying 4 *MMGs*, one *LMG* and innumerable Japs. One Jap threw a grenade at Lieut Swindell's tank which landed on the turret and exploded. Lieut Swindells was badly wounded and the tank was sent back, but he died on arrival in harbour. The Squadron withdrew to harbour East of *Impetlet* by 1730 hours. Capt Nicolson's tank was left with the infantry company which remained for the night at M. S. 4½. Ris Siri Singh's tank had been recovered during the afternoon.

'C' Squadron, which had been moving along the left axis, met little opposition. *Leindaw* and *Kyangyagon* were soon reported to be clear of enemy. The squadron later cleared the group of villages North East of *Leindaw* and harboured just South of the village.

128 Jap bodies were counted in and around *Impetlet*, and a further 50 or 60 were claimed in the area of M. S. 5 and *Subyugwet*. Three 47 mm anti tank and one 70 mm gun were dragged into harbour West of *Impetlet* that night. Throughout the day soft vehicles moving on the road to *Meiktila* had been shelled whenever they came into range of the Jap gunners. One shell landed close to the D 8 (recovery tractor) sent to tow in Capt Nicolson's tank, the driver being killed. Except for intermittent shelling during the night onto the edges of the harbour, a quiet night was spent after a very hard day.

The following day, the 18th March, 'A' Squadron was ordered to mop up the area South of *Myindawgan* between the two roads. 'C' Squadron was to carry out a wide sweep North and East of Point 827 with the object destroying all enemy in that area. Owing to the failure of 99 Indian Infantry Brigade to capture Point 801, the infantry with 'A' Squadron were pinned down by artillery fire and the force was unable to achieve its object. 'C' Squadron, however, had a most succeessful sweep, killing many enemy and destroying much equipment. Both squadrons harboured in their previous night's harbours.

On the 19th March two troops, one each from 'A' and 'C' Squadrons, under command of Capt Riazul Karim Khan, and one platoon 1/10 Gurkha Rifles formed up West of *Subyugwet* with the intention of clearing up the nullah area running South and East from *Myindawgon* Lake. The, advance reached the high ground South East of *Subyuget* but, immediately tanks and infantry appeared on the forward slope intense 105 mm gun fire broke up the advance. Capt Riazul Karim Khan picked up three wounded infantry and put them onto his tank. The attack was abandoned as it was obvious that no progress could be made until the enemy gun positions between M. S. 3 and M. S. 4 on the *Pindale* road had been neutralised. Later the whole withdrew to *Meiktila.'* Our casualties were light, but eight soft vehicles and two recovery vehicles had been destroyed. Enemy casualties were estimated as follows:—

264 killed.

2 105 mm guns destroyed.

4 anti tank guns destroyed.

2 truck and 1 jeep destroyed.

On the 20th March 'B' Squadron (Major Stewart) moved to 99 Indian Infantry Brigade harbour on the *Meiktila* airfield. The brigade less one battalion, with 'B' Squadron in support, had been ordered to clear the group of villages at *Nyaungbintha* and to exploit East as time allowed. Night patrols reported that no enemy were holding this village, so the following morning the 1/3 Gurkha Rifles were ordered to send two companies through it. 'B' Squadron was ordered to remain in harbour until called for. At 0900 hours two companies crossed the railway line West of the village, and an hour and a half later reported that enemy, strength one section, with an *LMG* had been seen. Shortly after they reported that consider-

AN 'A' SQUADRON CONFERENCE

L. to R:— Lt. A. W. LANE — CAPT. M. A. NICOLSON
Lt. BAHADUR SINGH — MAJ. B. L. LORAINE-SMITH
RIS. SIRI SINGH — RIS. BHAG SINGH

able opposition was being encountered. A troop of tanks was ordered forward, but before it arrived the infantry were ordered to withdraw West of the railway. The remainder of the squadron moved up to M. S. 2½. An airstrike, mortar and artillery concentrations were but down after which two companies of infantry, with 'B' Squadron less one troop in support, advanced into the village. One Troop (Lieut Carey), with one platoon 6/15 Punjab Regiment, had been ordered to move round the North of *Nyaungbintha* and to take up a position covering the East, which was a probable line of enemy withdrawal. Progress was very slow owing to the thickness of the trees and to the very wide front allotted to the squadron. It appeared that the enemy were moving round onto the flanks of the second wave of infantry. By 1630 hours our troops had arrived at within 400 yards of the far edge of the village, but it was obvious that the whole position could not be cleared before dark, so withdrawal was ordered to commence at 1650 hours. Meanwhile Lieut Carey's troop was being heavily sniped and mortared, and the infantry with it were sustanining casualties. With much difficulty these were all recovered by being lifted into the tanks through the escape hatches. Lieut Carey was wounded in the hand while directing a tank by means of the infantry call-box. His troop returned to the airfield across country and did not attempt to retrace its tracks onto the main road. The brigade claimed to have killed 168 enemy, a figure which was thought to be excessive. The squadron, however, claimed to have killed 40.

At 0900 hours on the 23rd March, the Regiment less one squadron, under command of 63 Indian Infantry Brigade, moved out of *Meiktila* with the intention of clearing the enemy in the *Kinde* and *Kandaung* area. These two villages are South of *Meiktila*, and had been attacked twice on preceeding days by the Royal Deccan Horse, but had not been cleared. It was believed that an unsuccessful attack on the perimeter of *Meiktila* on the night 21/22nd March had been carried out by a battalion based on *Kinde*. It was hoped that they would be in no form to resist a third attack after their unsuccessful night operation. The force moved out in the following order to *Nwazantaung*:—

Advance Guard.

Comd.	Lieut. Col. Smeeton.
Tps.	'A' and 'C' Sqns, Probyn's Horse.
	Two Coys, Border Regt.
	Det 36 Fd Sqn. I. E.

Main Body.

Tps	Border Regt., less two Coys 7/10 Baluch. A. D. S.
In support	Div Arty. 59/18 Bty, R. A.

The orders given to the advance guard commander were to recce *Kinde* and, if it was found to be occupied, to attack and clear it. The advance guard moved with 'A' Squadron (Major Loraine-Smith) leading to the high ground 500 yards West of *Kinde*, where both squadrons took up 'hull down' positions. One troop and one platoon were sent to recce the North-West edge of the pagoda area, whilst a second troop and platoon recced the North-West edge of *Kinde* itself. Both troops met with sniper opposition, and the left hand troop was fired at by a 75 mm gun. The two troops were called back with the intention of putting in an attack on the village. During the recces, a B.O.R. of the Borders sat down under a tree near the advance guard H. Q., and looking up saw five Japs. These were shot down like rooks. Shortly after, a sixth Jap was heard shouting, 'Master, master' from the same tree. He was brought down alive, but was later shot whilst trying to escape. The attack was planned to go in to *Kinde* on a two squadron front with the East end of the village as the objective. The left hand troop of tanks, which had been doing the initial recce, was very slow in withdrawing because its infantry lost one of their wounded. When the troop leader finally reported, he said he was convinced that the 75 mm gun had fired on him from the East end of the pagoda area. The plan was therefore modified, and one squadron with one company of infantry was directed onto the pagoda area, one squadron and one company onto the North side of *Kinde*, and the third company of the Border Regt. which had then arrived at the advance guard H. Q., onto the Southern edge of the village. An ineffective airstrike was put onto the pagoda area, the majority of the bombs failing to explode. A concentration was put down by the divisional artillery onto the village, during which the tanks moved to within 100 yards of the West edge. After the concentration the tanks had a five minutes' shoot, and then continuing their fire moved to the fringe of the village. The infantry, who were following closely, passed through the tanks into the village and pagoda area. The right company was immediately engaged from bunker and dug-in fire positions. The left company made good progress but was held up on the right by machine gun fire coming from the North end of the village. The usual unlocated sniping from trees was continuing the whole time. The tanks then moved up to the infantry. Slow but steady progress was made, and the whole of the pagoda area was finally cleared. The toughest fighting took place in the centre where Capt. O'Beirne-Ryan's tank, leading the right hand troop, was hit by a hollow-charge shell and penetrated. The driver and lap gunner were both wounded and the tank caught fire. This was extinguished and the tank was driven out of action by Capt. O'Beirne-Ryan himself. This officer, after given a report of the action to Lieut. Col. Smeeton, then went back on foot into the village to find the wounded lap gunner. He was killed by a grenade in the thick of the fighting. The lap gunner, unknown to this officer, having been already evacuated to the A. D. S. The leading tank of Lt. Elder's troop on reaching the place where Capt. O'Beirne-Ryan's tank had been hit was also set on fire by a hollow-charge shell.

The advance onto the North side of *Kinde* village progressed, and a troop was sent from the pagoda area down to the North edge of the village to assist the advance. One tank of this troop was also hit by a hollow-charge shell and caught fire. It was driven out of action guided by the tank commander on foot; flames were coming out of the turret. As soon as it was clear of the village, fire extingu shers were used and the tank was closed. It continued to mioulder and later blew up. It became obvious that the village could not be cleared by dark, and the Brigadier ordered the attack to be called off, if the situation had not cleared, with in fifteen minutes. After that time the infantry were ordered to withdraw. Lt. Elder had fixed a tow-chain to the burning tank in an attempt to recover it, but on the way out of the village the tank struck a tree and the tow chain broke. He was then ordered to leave it, as by then flames were coming out of the turret; shortly afterwards the tank blew up. The force reached harbour, one mile North-West of *Kinde*, without further incident. During the night three platoons of enemy were seen to leave *Kinde* and approach the harbour, but they turned North before reaching the perimeter. Enemy vehicles were also heard moving into *Kinde.*

The following day, 24th March, 'A' Squadron was ordered to recce *Kandaung* village, and inflict as much damage as possible without becoming engaged in a major battle. The squadron moved off at 0700 hours, passing North of *Kinde* to *Kantha*, which village they reported clear, and then moving down the road to *Taingon*. Here a small party of enemy was encountered and engaged, and the villages on either side of the road cleared. The squadron pushed on to *Kandaung*, and patrols were sent into the village. These were fired on from the pagoda area and withdrew. An airstrike was put in, the tanks then moving forward to the East side of the village, with one troop attacking the North-East corner. Enemy were observed in the pagoda area and engaged. After a tank shoot on the village and pagoda area, patrols were again sent in, only to be met with further resistance. The squadron was then ordered to withdraw.

During the morning, 'tiger' patrols of the Baluch Regiment had entered *Kinde* and reported it clear. 'C' Squadron was moved down to the high ground West of the village, and two companies of the Baluch were ordered to sweep it supported by 'C' Squadron. 'Tiger' patrols were sent in a second time to confirm that the village was clear, but on this occasion they were engaged from what appeared to be a platoon position on the North edge of the village. The Brigade commander decided to occupy the pagoda area, which was not held, with a company of the Borders, and to put in a company attack, supported by 'C' Squadron, onto the platoon position. The plan for the attack was in fact exactly the same as on the previous day, except that a platoon was put in on the right instead of a company, and that instead of an artillery concentration, the combined mortars of the Baluch and Border Regiments were to put down a barrage before the tanks' and infantry's advance. This plan was put into operation, but as the infantry advanced behind the tanks, a 75 mm shell fell onto the infantry company's H. Q. wounding the company commander and about ten men. Some time was spent in reorganization before the infantry were ready to follow the tanks. The advance continued, but the infantry did not seem to be clear as to their objective, and were not well dispersed. Although the right troop of tanks went into the North end of the village, the majority of the infantry followed the left troop between the pagoda and the village, and were very quickly held up by fire from the village. The Borders in the pagoda area were also engaged from the same position and suffered some casualties. The left troop pushed forward along the North edge of the village, but it was some time before the infantry would follow. Just as they started to advance, a 75 mm shell hit the leading tank, smashing both the driver's and lap gunner's periscopes, and so distorting the frames that they could not be replaced. The tank reversed and a second tank went forward in its place, but the infantry withdrew to the position that they had just left, where they found some cover behind a bund. A second tank was hit by a hollow-change shell. All the turret crew were wounded, a fire started, and the crew evacuated the tank under heavy small arms fire without further casualties. The troop leader moved close up to the burning tank, and Dfr. Piara Singh dismounted and used his fire extinguisher but with no effect. He then dragged the blazing tarpaulin off the tank, which fell under the engine doors. The troop leader asked permission to fix a tow chain, but as he was quite unsupported by infantry and within twenty yards of the enemy's position, he was ordered by the C. O. to withdraw, since it was considered that the tank would subsequently blow up. It was also not considered worth the risk of losing another tank to this, as yet unlocated, hollow-charge weapon.

The Brigade commander decided to call off the attack, and the infantry withdrew covered by the tanks. On this occasion, however, the withdrawal was followed by enemy snipers working wide on the flanks who used grenade dischargers. The force moved back to *Thayetkon*, where the night was spent.

On further enquiry from Dfr. Piara Singh and the troop leader, it appeared that, apart from the local fire in the turret of the abandoned tank, only the clothing bin and tarpaulin caught fire; it was therefore considered possible that the tank had not burnt out completely. 'C' Squadron was therefore sent back the next morning to recce the village and to report on the condition of the tank. On arrival, they found that the tank had been moved by the enemy. It had now become necessary to clear the North of *Meiktila*, and the area of the airfield, so this force was ordered back for the operation, the tank having been left in enemy hands. That night it was driven up to the primeter wire by the enemy but, with complete loyalty, it turned itself upside down in a nullah and burnt itself out.

On the night of 24/25 March, considerable enemy activity was observed in the area of the harbour at the *Khanda* level crossing and also at M. S. 339 on the *Mandalav* road. Consequently 'B' Squadron (Major Stewart) was ordered to proceed to the area South of M. S. 339 and 'marry up' with one company of the West Yorkshire Regiment for an operation commanded by 48 Indian Infantry Brigade. Infantry patrols and reported that *Kyigon* was held. No 3 Troop (Jem Gurbachan Singh) with one platoon was ordered up to the pagodas on the outskirts of the village and then to go through it. Soon after this troop with its infantry had formed np, shelling from Jap mortars 75 mm and 105 mm guns commenced. The infantry were pinned to the ground and sustained casualties. The troop advanced and engaged the enemy destroying a 75 mm gun. No 3 Troop was then ordered to remain in position while No 2 Troop (Lieut Grover) with one platoon, under command of Capt Udham Singh, was ordered to move forward on the left of No 3 Troop. When No 2 Troop had advanced its infantry were also pinned to the ground. The tanks, therefore, moved forward locating and destroying enemy guns. Jem Gurbachan Singh destroyed a 70 mm gun and Lieut Grover a 75 mm gun. The enemy was well dug in and the advance was very slow. Squadron H.Q. moved up with No 1 Troop (Ris Ujagar Singh) and another company of the Yest Yorkshire Regiment. By degrees the tanks and intantry cleared the area. Altogether the following were destroyed:—

1 75 mm gun.	2 morters.
1 70 mm gun.	3 *MMGs.*
1 47 mm gun.	40 killed.

The major half of this action was commanded by Capt Udhan Singh, who kept excellent control in very thick country. He was under shell fire during most of the day which did not make his task easy.

The 26th March was spent in harbour, and on the 27th the Regiment, under command of Tac H.Q. 255 Indian Tank Brigade, formed part of a force known as *'Guncol'* which made a sweep round the North of *Meiktila* from West to East, in an attempt to destroy enemy gun positions. The force harboured for the night at *Kyaukpyugon.* The following day a similar sweep was undertaken and a few enemy gun positions were located and attacked. Slight opposition was encountered and the force harboured in its previous night's camp. On the 29th March the whole area was cleared of enemy, and *'Guncol'* returned to permanent harbour at 1600 hours.

CHAPTER III.

Chapter 3 of the story of the operations carried out by Probyn's Horse in *Burma* tells of the advance from *Meiktila* Southwards down the railway corridor from *Meiktila* to *Hlegu*, 32 miles North of *Rangoon*. The total distance excluding 'hooks' and diversions being over 300 miles, the whole advance being completed in 30 days.

In the beginning of April 1945, 255 Indian Tank Brigade was on the West bank of *Meiktila* Lake (South). The Brigade had just finished a month of individual actions around the town itself.

B 2 and B3 Echelons, which had been separated since the *Irrawaddy* Crossing, arrived safely, and squadron commanders were particularly pleased at the arrival of their spare crews, to say nothing of their langar and mess vehicles.

The first phase of the advance was the capture of *Pyawbwe* by 17 Indian Division with under command 255 Indian Tank Brigade. The plan was briefly as follows:—

99 Indian Infantry Brigade was to move out East from *Meiktila* to *Thazi*, and thence South to capture Pt 825, which lies about 3 miles North of *Pyawbwe*. 48 Indian Infantry Brigade was to move South down the main road, take *Yindaw* and hold a firm base there while *'Claudcol'* and 63 Indian Infantry Brigade were to proceed South from *Yindaw* to *Yenaung*, where 63Brigade was to turn East and attack *Pyawbwe* from that direction. *'Claucdol'* was to proceed South to *Ywadan*, then North East to cut the main *Rangoon* road South *Pyawbwe*.

The composition of *Claudcol* was as follows:—

Tac Bde HQ.
Probyn's Horse.
One Sqn 11 Cav. (P.A.V.O.)
6/7 Rajputs.
4/4 Bombay Grs less three coys.
7/10 Baluch.
36 Fd Sqn. I. E. less two tps.
59/18 Bty R. A.

On the 4th April, the column moved to *Kandaung* which is South of *Meiklila*. All ranks were delighted to be on the move again, as *Meiktila* had become very hot and the flies were most troublesome.

On the 5th April, 48 Indian Infantry Brigade moved down the road brushing aside minor opposition until *Yindaw* was reached. It soon became apparent that the village was strongly held, so it was decided that *'Claudcol'* should take over the attack.

Yindaw is bounded on the west by a large jheel, and on the other three sides by a moat and a high thickly wooded bund. There was no time for a preliminary reconnaissance to be made but, from information received, it was concluded that tanks could only enter on the North side where two causeways passed over the moat and through the bund, these being anti-tank obstacles. The attack was planned in three phases:—

Phase 1. Half 'A' Squadron, under command Major Loraine-Smith, was to support 'A' Company 6/7 Rajputs onto a road running East to West parallel to and North of the village.

Phase 2. 'D' Company 6/7 Rajputs was to pass through 'A' Company and secure a bridgehead to include one or both causeways, followed by sappers and the first half squadron of tanks. It was thought certain that the causeways would be mined.

Phase 3. The second half of 'A' Squadron, under command of Capt Nicolson, was to pass over the causeway and clear the village inside. Meanwhile 'C' Squadron was to take up fire support positions East of the village. 'B' Squadron was to remain in reserve.

The first half of 'A' Squadron, with No. 2 Troop on the left and No 1 Troop on the right, left the F. U. P. one mile North of *Yindaw* at 1415 hours, and Phase 1 was easily completed with only slight sniper opposition. An artillery concentration was then put down immediately South of the bund, strengthened by prophylactic fire from the tanks. 'D' Company 6/7 Rajputs advanced and the sappers began to work towards the left causeway. Enemy opposition over the bund was very strong indeed and, inspite of the most gallant efforts, the infantry were driven back. The sappers lifted one large mine from the left causeway before they were compelled to retire. Owing to the height of the bund the tanks were unable to see and destroy the enemy positions Both causeways were blocked by felled trees and, owing to enemy fire, the sappers were unable to approach the right causeway. In order to support the infantry No. 1 Troop (Lt. Lane) was ordered to work along the left causeway and Jem. Sarup Singh took his tank right up to the road block, which he blasted but was unable to move. He also engaged all targets which he could see inside the village. 'B' Company 6/7 Rajputs was next called up, and two attempts were made to get over the bund, but without success. The infantry were unable to move forward as they were suffering very heavy casualties and, as the tanks could not get into the village to give close fire support from outside, it was decided to withdraw. The second half squadron had been ordered up and were in a position echeloned to the left and rear. The withdrawal took a long time, as there were many casualties to be evacuated and enemy sniping became intense. It was not possible to bring back the infantry casualties from inside the village, but all other casualties were brought back. The Squadron covered the withdrawal as follows, by airburst shells in the trees inside the village and intensive fire onto and over the top of the bund to prevent the enemy from following up:-

The first half of 'A' Squadron followed at the rear of the infantry, followed by 'C' Squadron, followed by the second half of 'A' Squadon. The fire support given by 'C' Squadron was much hampered by thick trees and the height of the bund. The Squadrons harboured two miles North of *Yindaw* with the remainder of the Regiment at 1900 hours. Enemy casualties were estimated at 60 killed.

On the 6th April, the Regiment, under the command of of *'Claudcol'*, moved from *Ywathit* to *Kanbya* bye-passing *Yindaw*, which was subsequently evacuated some days later. Defences were found to be extensive and extremely well dug. The distance to *Kanbya* was very short, but the Regiment experienced difficulty in getting into day dispersal areas owing to a deep sandy nullah which soft vehicles were unable to negotiate until work had been done to it. 'B' Squadron's area contained a large onion field and a well, both of which were greatly appreciated.

On the 7th April, 'B' Squadron was included in the advance guard to *'Claudcol'* while 'A' and 'C' Squadrons moved in the main body. The commander was Lieut-Col. Tighe

4/4 Bombay Grenadiers. During the first part of the day the advance continued Southwards without opposition but, on reaching a large nullah North of *Yawdin*, opposition was encountered actually in the nullah and on the far bank. The plan to clear it was as follows:—

No. 3 Troop (Ris Gurbachan Singh) and No. 2 Troop (Ris Ujagar Singh) under the Squadron 2nd-in-command (Major Udham Singh) was to attack the high ground on the South side of the nullah, and on the East side where there was known to be an enemy position.

This attack was preceeded by an artillery concentration from 59/18 Bty R.A. and ,on completion, the Bombay Grenadiers with two platoons attacked the feature supported by two troops of tanks. It was successfully taken, and 2 LMGs were captured with 14 Japs killed including one officer. In order to save time while the attack was going on No 2 Troop (Lt. Grover) and No 4 Troop (Capt. Mustafa Khan) were sent to investigate the village of *Ywadin*, which was found to be clear. Major Udham Singh, on completion of the hill battle, rallied back his force and the advance continued with No. 2. Troop leading.

About two miles further down the road the leading tank (Jem Sikander Singh) came to a nullah, whose blown bridge was an anti tank obstacle. The leading tank attempted to cross but got badly stuck and shed its track trying to get out. The only hope of getting across was to construct a diversion to the left of the bridge. Sappers were called up and work started. The Japs soon began to counterattack from very thick scrub on the right of the bridge, and work stopped temporarily. The situation, however, was quickly restored by the infantry with fire support from No 4 Troop and work was resumed. The Japs continued to give trouble from the right of the road but not in sufficient force to prevent a crossing being made, which was completed in about one hour.

No 2 and No 3 Troops continued the advance, having crossed the nullah under Major Udham Singh, and carried on several miles until they came under heavy fire from 47 mm and 75mm guns. It was now about 1730 hours and so the force was ordered to withdraw back to harbour at *Ywadin*. No 4 and No 1 Troops had in the meantime been dealing with the situation round the diversion which was still not completely

A REGIMENTAL CONFERENCE

L. to R: – MAJOR W. M. ARKINSTALL — MAJOR F. W. KENNEDY
CAPT. E. HALLIWELL — MAJOR B. L. LORAINE-SMITH —
Lt. Col. M.R. SMEETON AND TWO INFANTRY OFFICERS.

clear and, when they finally withdrew to harbour at *Ywadin*, an ambush of 14 men was left by the 4/4 Bombay Grenadiers on the diversion, but they were forced to withdraw by superior enemy forces as soon as it got dark. Jem Sikander Singh's tank was towed back to harbour by a D. 8 (recovery vehicle) and the track was collected the following morning.

On the 8th April, the advance was resumed and the nullah from which the infantry patrol had been driven back on the previous evening was found to have been deserted. At first all went well. 'A' Squadron formed the advance guard with Lieut Bahadur Singh's troop, which carried no infantry. At about 0715 hourswhen 400 yards North of *Yenaung*, anti-tank fire opened up from the direction of the bund on to No. 4 Troop which was advancing down the axis of the road and to the East. Two tanks were hit and the troop withdrew to a hull down position about 300 yards in rear and engaged the anti-tank gun. Squadron HQ. moved up on the West of the road level with No. 4 Troop, and 59/18 Bty R.A. put down a concentration on the corner of the bund and knocked out the gun (47 mm).

The C. O. (Lieut-Col Smeeton) then came up and a plan was formed with the intention of capturing *Yanaung*. 'A' Squadron was to clear the arca of the bund and cross-roads, and 'C' Squadron was to move in a right hook to the West and South. During this time small arms fire was coming from the scrub South of the bund, and enemy infantry could also be seen who were engaged by Squadron H. Q. and No. 4 Troop.

After troop shoots by No. 4 Troop (Lieut. Bahadur Singh) Squadron H. Q. and No. 2 Troop (Risaldar Siri Singh) on the cross-roads and scrub area, No. 2 Troop moved to the end of the bund and, having dismounted the infantry, began working up the far side. This manoeuvre produced an immediate effect, and the Japs began to filter back across the open ground towards the village, thus providing easy targets. No. 3 Troop (Risaldar Bhag Singh) was immediately ordered up on the right of No. 2 Troop, and Squadron H. Q. moved to the end of the bund.

The original appreciation that the Japs were holding the ground in front of the village was now confirmed. *Yanaung* itself was completely burnt out as the result of an air-strike the evening before. Enemy positions were seen to be all along the bund in a dry nullah and in the area of the cross-roads and to the East.

When No. 4 Troop started to move forward to cut the *Pyawbwe* road, it was engaged by two 47 mm anti-tank guns sited by the side of the road. Fire was returned, and during a short action one enemy gun was knocked out; Lieut. Gilkison's tank was also damaged by direct hits. No. 4 Troop then advanced up the road, and the Japs were seen to be taking their second gun into a village some 800 yards to the East. They were engaged and the gun was later captured by "B" Squadron when they attacked that village.

Meanwhile No. 1 Troop (Lieut. Lane) had been ordered up to clear the scrub area between the bund and the dry nullah. The Japs here were very numerous and required a lot of "winkling out" of their foxholes. By the bend in the bund the positions were strongest and took some time to clear. The 2nd-in-command (Capt. Nicolson) with a section of infantry, worked up the nullah on the left of No. 1 Troop, followed by No. 3 Troop on his left, which crossed the road and cleared the dry nullah area. During this time the Japs were attempting to run back into the village, and great slaughter was done West of the jheel and around the cross-roads.

After the bund areas had been cleared up to the road, it was seen that the remaining enemy had retreated into scrub enclosed by a small bund North of the *Pyawbwe* road. No. 3 Troop advanced to give fire support. No. 1 Troop crossed the main road to clear the area from West to East, No. 4 Troop was called back into reserve. To clear this area took some time, and the squadron then rallied in the area of the cross-roads while mines and a road block were lifted. Any Japs who had escaped retreated straight through the village and on Southwards, but over 150 dead were counted on the ground after the completion of the action.

Meanwhile No. 4 Troop of 'C' Squadron (Lieut. Elder) had cut the road South of the village and, with one platoon of infantry, remained in position until relieved by armoured cars. No. 2 Troop (Lieut. Robertson) had made a sweep South of the village and destroyed 15 Japs who were trying to escape. No. 2 Troop and No. 3 Troop (Risaldar Mohd Arif Khan) probed into the South-West of the village and cleared it. Heavy fires made advance impossible. No. 2 Troop was later ordered to make a right sweep through the village with two companies 6/7 Rajputs, the tanks keeping on the main road; this was completed without event.

The advance South was continued at 1500 hours with armoured cars leading followed by No. 4 Troop 'A' Squadron. They arrived at within 400 yards of *Ywadan* but were called back to harbour about 7 miles South of *Yanaung*.

During this action the Regiment sustained :—

One I.O.R. killed.

One I.O.R. wounded

Enemy casualties were :—

230 killed
3 47 mm Anti-Tank guns destroyed.
4 M. M. Gs. destroyed.
3 M. M. Gs. captured.
5 L. M. Gs. destroyed.
4 L. M. Gs. captured.
20 rifles captured.

At about 1930 hours one Jap tank moving North on the road towards the harbour was stopped by our own Sappers from running over our own mines laid in the road. It turned about and disappeared into the darkness after firing one round at the harbour which did no damage.

On the 8th April, when 'A' and 'C' Squadrons moved off to *Ywadan*, 'B' Squadron came under command of 63 Indian Infantry Brigade. Half the squadron consisting of No. 3 Troop (Risaldar Gurbachan Singh) and No. 4 Troop (Capt. Mustafa Khan), commanded by Major Stewart, was ordered to attack a village about one mile North East of *Ywadan* and then to capture the high ground at Pt. 900, which was on the far side of it. No. 3 and No. 4 Troops, having dismounted their infantry short of the village, went through it with the infantry leading and the tanks in support: No. 4 Troop left and No. 3 Troop right. The village was difficult to get through as it was burning hard, the result of an air-strike firing the thick vegetation. On reaching the far side, the village was found to be clear and No. 3 Troop destroyed an enemy 47 mm gun which was in the open on the right of the village.

The approach up to Pt. 300 necessitated the crossing of a nullah which was an anti-tank obstacle to No. 4 Troop. No. 4 Troop then came in behind No 3 Troop and continued to advance up the hill. No opposition was met until both troops were practically on the feature. No. 4 Troop then spotted some enemy in a large nullah and engaged them killing quite a few, while No. 3 Troop consolidated the top of the hill. No. 4 Troop later joined No. 3 Troop and a position of all round defence with the infantry was taken up until the arrival of the Baluch Regiment, who had put in a battalion attack astnde the road from the North. During the consolidation period innumerable Japs tried to run away and provided excellent shooting for the tanks. The F. O. O. with the squadron put down a very good shoot on a suspected position South of Pt. 900, and No. 4 Troop supported it.

The tanks then withdrew from the feature and, having met up with the remainder of the squadron, went down the *Yanaung—Pyawbwe* road where they harboured for the night with 63 Indian Infantry Brigade at the village of *Kyauktang*.

On the 9th April, the advance was continued down the *Yanaung—Yamethin* road, and the intention of the armoured column was to pass through *Ywadan*, cross the chaung and to turn North East cutting the *Pyawbwe—Yamethin* road in the Pagoda area at M. S. 305 on the main *Rangoon* road. According to local reports *Ywadan* was clear of the enemy.

'A' Squadron continued as advance guard and, in the wake of a troop of armoured cars, moved across country to the South of *Letha*, which was cleared by No. 1 Troop (Lieut Lane). No. 4 Troop (Lieut Bahadur Singh) bye-passing *Ywadan* reached the nullah bank where it destroyed a 75 mm gun sited under a large tree on the East side of the main road and nullah crossing. The enemy did not fire a shot, but the gun crew and the protecting infantry were seen withdrawing into a nearby villaged and were engaged. Meanwhile No. 2 Troop (Risaldar Siri Singh), who had been doing right flank protection, moved up and together with infantry cleared the village from North to South. It appeared that the enemy was formed of untrained units who did not fight with the usual

Jap tenacity. The nullah proved to be a considerable obstacle.

On local information that *Sedeik* was the main enemy position, a concentration was put down by both medium and S. P. guns, and the village was set ablaze. No. 4 Troop followed by No. 3 Troop (Risaldar Bhag Singh) then crossed the nullah preparatory to clearing *Sedeik* from North to South with No. 4 troop acting was a 'stop'. As the tanks were getting into position they were engaged by an enemy gun of large calibre from the Southern end of *Okpo Ashe*. Fire was returned and the gun was knocked out by the Squadron 2nd-in-command (Capt. Nicolson) shooting across the nullah. No. 4 Troop then carried on and killed enemy who were retreating down the main road, while No. 3 Troop cleared the village and accounted for a few enemy still hiding in their bunkers. At the same time No. 2 Troop followed by No. 1 Troop and Squadron. H. Q. crossed the chaung by the main crossing after it had been swept by the sappers. No. 2 Troop came into action immediately East and South of the crossing against enemy bunkers and foxholes. As it was then obvious that the villages on the South bank of the chaung, South of the main road crossing, were held, 'C' Squadron was ordered up to clear them. Having crossed the nullah, several large explosions took place near squadron H. Q.; these were afterwards found to be electrically detonated mines. One Jap was lying on the top of a mine, shamming death, but moved slightly and was shot; the mine underneath him exploded. On the edge of the village near the nullah several Japs in bunkers resisted for some time even after nearby buildings had been set on fire. No. 4 Troop advanced South clearing slight opposition. No. 2 Troop cleared the nullah area and the whole Squadron moved South. Approximately two hundred Japs were seen trying to get away, a large number being killed crossing a patch of open ground. Jap casualties counted were fifty dead, our own nil.

Meanwhile No. 1 Troop of 'A' Squadron, (Lt. Lane), moved up to *Okpo Ashe*, which was definitely held, and another artillery concentration was put down with excellent results. The infantry then went through and, after mopping up, thirty enemy dead were counted and two abandoned 105 mm guns were found, one of which had been previously knocked out by No. 1 Troop, and the other damaged. It was surrounded by its dead crew. No. 2 Troop which had a good kill, having been relieved by 'C' Squadron, was moved up to clear *Letpankahla*. 'A' Squadron then continued the advance North-East to the high ground in the pagoda area at m.s. 305 South of *Pyawbwe*, there the force harbouring for the night.

As the result of local reports one platoon of infantry with one troop 'C' Squadron (Risaldar Sharaf Ali Khan) went to *Gwegon* village and there found 17 hastily abandoned Jap trucks. These were methodically destroyed by Lieut-Colonel Smeeton.

At approximately 2030 hours that evening, a Jap convoy using headlights approached the harbour from *Yamethin*. This was engaged by 'C' Squadron, whose tanks were harboured parallel to and 100 yards from the road. On orders from the C. O. the word to fire was given as the convoy came level with the tanks. Cries of dismay could be heard from the surprised Japs! The next morning three enemy lorries were found burnt out, and two more were abandoned. Later the same evening three tanks approached the harbour from the direction of *Pyawbwe*, the Japs being apparently still unaware of our presence in the area. One tank was knocked out by 'C' Squadron's 75 mm gun fire, one abandoned after it had left the road and over turned in a nullah outside the North edge of our perimeter, and the third after turning about, retraced its steps for 1½ miles, overturned and burnt out. It was claimed by Medium and S. P. Gunners who had D. F. tasks in that area.

Our casualties that day were two V. C. Os. and two I. O. Rs. wounded. The enemy casualties were:—

220 Killed
2 wounded
2 150 mm guns destroyed
3 medium tanks destroyed
1 75 mm gun destroyed
26 trucks destroyed

After the nullah at *Ywadan* had been crossed, a force consisting of the 4/4 Bombay Grenadiers less three companies and one link troop of armoured cars, (16 Cavalry) carried out recce in force down the road *Ywadan—Yamethin*. The following day *Yamethin* was entered, 30 enemy were killed and 3 vehicles destroyed. The town appeared to be only lightly held. As the force was not strong enough and could not be reinforced easily, it was ordered to withdraw to the Brigade harbour south of *Pyawbwe.* The fact that there were not enough troops to hold *Yamethin* had serious repercussions later.

On the 9th April, 'B' Squadron was still under command of 63 Indian Infantry Brigade. Half the squadron under the Squadron 2nd-in-command (Capt. Udham Singh) was ordered to carry out a recce in force on the South West of *Pyawbwe* with a view to an attack on the following day. No. 1 Troop (Jem Bhag Singh) and No. 2 Troop (Lieut. Grover) "married up" with two platoons of 'C' Company the Border Regiment with the object of pushing into *Pyawbwe.*

No. 1 Troop advanced from the start line with one platoon in the formation "two up": Dfr Tarlochan Singh's tank on the left, Jem Bhag Singh's tank on the right, and Dfr Harnam Singh's in near. When the troop had advanced nearly 300 yards a Jap ran away from the I. B. Canal. The infantry were immediately dismounted. When they were 100 yards short of the Canal four 2 inch mortars fired at Dfr Tarlochan Singh's tank. The troop halted and put down a troop shoot onto a bunker position 200 yards to their front. During this time one 105 mm gun and one 75 mm gun were firing from the main road onto the rear of the troop After the troop shoot No. 1 Troop advanced up to the I. B. Canal and Dfr Tarlochan Singh's tank crossed the nullah covered by the other two tanks. As it was crossing, enemy 2 inch mortars again opened up. Jem Bhag Singh himself went forward and crossed without mishap. The troop destroyed all the bunkers and killed many Japs who tried to run away. Later in the day the troop received orders to go forward without infantry. While doing so, Jem Bhag Singh destroyed an enemy battalion gun. Dfr Tarlochan Singh's tank had its ammunition box, behind the driver, hit by a 4 inch mortar bomb. Although the armour was pierced no damage was done. Later the same tank was again hit by a hollow charge shell which penetrated but luckily hit the water containers first. One 75 mm A. P. shell hit the turret but failed to penetrate it and yet another 37 mm shell shot away two track connectors. Orders were eventually given for the withdrawal of No. 1 Troop. Jem. Bhag Singh was wounded by a 75 mm shell which hit his cupola as he was withdrawing.

No. 2 Troop (Lieut. Grover) with one platoon of infantry had been ordered to clear the village on the *Thybin* canal and then find a crossing on the I. B. canal and probe forward to the railway line. The whole area was found to be clear, but on reaching the Canal, the troop was heavily shelled by a 105 mm gun. After crossing, small arms and 70 mm fire came from some long grass. This was silenced. The troop leader located and destroyed a battalion gun on the high ground near the railway line. After that the troop advanced to the railway line and was heavily sniped. Lieut Grover's tank stuck and while fixing the tow-rope, ALD. Harcharn Singh was wounded, also Jem. Sikandar Singh was hit in the shoulder. The tank was pulled out and sent back with Jem. Sikandar Singh still commanding it. A.L.D. Harcharn Singh was carried back on Sowar Kishan Singh's shoulder for a distance of 300 yards then put onto Jem. Sikandar Singh's tank. By this time it was getting late and orders were received for tanks and infantry to withdraw to 63 Indian Infantry Brigade harbour.

On the 10th April, *Claudcol* advanced North up the main *Rangoon* road towards *Pyawbwe.* 'C' Squadron was advance guard while 'A' Squadron moved with the main-body. Large the enemy dumps were destroyed in the villages of *Tabetka* and *Sibin.* The bridge over the *Thitson Chaung* was found to be blown and the enemy were holding the North and South banks. 'C' Squadron H. Q. forced a crossing West of the bridge and destroyed a Jap tank on the right of the road. No. 4 Troop (Lieut. Elder) crossed on the East and destroyed two medium tanks, which were being used as pill-boxes on the North bank of the chaung. It is interesting to note that no attempt whatsoever was made to manoeuvre them in any direction. Thirty-nine staff cars and trucks were destroyed or captured. 'C' Squadron pushed on to the road and railway crossing North of the chaung where sniping was again encountered but quickly dealt with. A large ammunition dump in the village of *Zindaw weir* was destroyed. No. 2 Troop (Lieut Robertson) was sent back to deal with a party Japs who had started sniping Brigade H. Q. which was in the main body of the column. 'A' Squadron then moved up the West side of the road and cleared the villages up to *Magyibin.* 'C' Squadron cleared the villages on the East of the road capturing a large sugar dump in the village of *Mindaw.* Both Squadrons advanced to within 600 yards of *Pyawbwe* and shot up a few enemy withdrawing Eastwards from the centre of the town. The column harboured two miles South of *Pyawbwe.* Considerable consternation was caused during the night among the infantry on the perimeter, because a party of Japs walked up to the wire, several of them being killed.

Enemy casualties were :—

100 killed	1 70 mm gun destroyed
1 105 mm gun destroyed	1 M. M. G. destroyed
2 75 mm guns destroyed	2 L. M. Gs. destroyed
2 37 mm guns destroyed	3 Medium tanks destroyed
10 Ammunition dumps destroyed	39 Vehicles destroyed or captured.

On the 10th April, 'B' Squadron, still with 63 Indian Infantry Bridge, were ordered to attack *Pyawbwe* from the West with two companies of the Border Regiment. The orders were as follows:—

The Squadron would move from its harbour area at the village of *Kyanktang* and be in position on the starting line by 0700 hours where it would marry up with 'A' and 'B' companies of the Border Regiment. No. 2 Troop (Lieut Grover) was allotted to the right company. No. 3 Troop (Jemadar Gurbachan Singh) to the left company. Captain Udham Singh with No. 1 Troop (Risaldar Ujagar Singh) and No. 4 Troop (Capt. Mustafa Khan) remained behind in the village of *Npeddaw*.

An artillary barrage was put onto the objective from 0723 to 0734 hours, and during this time the infantry supported by two troops of tanks moved up to the line of the I. B. Canal from which place the attack went in as soon as the barrage had finished. The squadron 2nd-in-command (Capt Udham Singh) had received orders to move up as soon as the barrage had started, and the squadron commander (Major Stewart) ordered him to take up a fire position North of the road and East of the I. B. Canal in order that he might support the attack going in on the South of the road.

All troops crossed the I. B. Canal safely, in spite of its having been mined. No. 2 and No. 3 Troops kept well up giving the infantry really close support. Progress was slow as the enemy was well dug in and possessed the odd L. M. G. which held up the infantry until they could be located. Meanwhile Captain Udham Singh's advance was progressing well and he was killing a number of Japs. Jemadar Nazar Singh knocked out a 75 mm gun and Captain Udham Singh a 37 mm gun.

On the South side of the road No. 3 Troop was progressing well and the company of the Border Regiment was making continued progress. No. 2 Troop was slightly held up owing to a few strong bunker positions which were in front of it, and was unable to advance. When No. 3 Troop was only 300 yards from the railway line, the leading infantry platoon was again pinned, and only after much supporting fire had been put down was the position cleared. By this time No. 2 Troop had managed to get forward again and was on the line of the railway. The squadron commander gave orders to No. 2 Troop to "hot up" the area for two minutes and then put its green flag as a signal to the infantry to advance. This was successfully done and No. 3 Troop with its infantry reached the line of the railway. Major Stewart ordered Captain Udham Singh to come down from the North to cross the railway, and to move down the main road to join his HQ. This move was successfully carried out except that Risaldar Ujagar Singh's tank was blown up on a mine and had to be towed away under heavy enemy shell fire.

Meanwhile No. 3 Troop had established itself on the high ridge East of the railway line and, while engaging enemy bunker positions, came under heavy shelling which forced it to pull back. As soon as it ceased, Lieut Grover brought his troop (No. 2) forward again and started to engage more enemy. While doing this his tank received a direct hit on the top of the turret from a 105 mm shell, which killed him and blew off the cupola ring. This troop had to be withdrawn as there was only one tank left in action, Dfr Balbir Singh's tank having broken down earlier in the day.

To the immediate front of No. 3 Troop, across the road, was an enemy position which was heavily engaged by tanks and infantry. It was not attacked as the squadron had reached its objective, but the next day a considerable number of dead were found in the area.

Enemy guns destroyed by 'B' Squadron were:—

2 75 mm

1 70 mm

CHAUNG CROSSING SOUTH OF TAUNGOO.

1 37 mm
1 M. M. G.
2 L. M. Gs.

Our total casaulties throughout the day were:—

one B. O. killed (Lieut K. K. Grover)
one VCO wounded (Jemadar Sikandar Singh)
one IOR wounded (ALD Harcharan Singh)

On the 11th April, the 6/7 Rajputs supported by 'A' and 'C' Squadrons were ordered to clear the South East quarter of *Pyawbwe.* The area was divided into three sectors for the purpose of the attack. Except for 3 Japs the town was found to be unoccupied. The enemy had laid mines indiscriminately along the roads and verges which made the advance of tanks through the town impracticable. Major W. M. Arkinstall, O. C. 'C' Squadron, returning from the new harbour location The Government Stud Farm North of the town, to guide his squadron into harbour ran over a mine when in his jeep and was killed outright. His driver Sowar Abdul Rehman sustained fatal injuries. Mines were also found in the harbour area, but these were removed without incident.

After the capture of *Pyawbwe,* 5 Indian Division took over the lead from 17 Indian Division. The advance guard was composed as follows:—

116 R. A. C. (Gordon Highlanders) less one sqn.
7th Light Cavalry (with under comd. one sqn. 16th Cav).

18th Fd. Regt. less one bty.

3/9 Jats.

One bty. (5th Mahratta) Anti-Tank.

A. D. S.

36 Fd. sqn. less one tp.

The column started on the 11th April but was delayed for two days while *Yamethin* was cleared. This town had been reported previously as lightly held. On the 14th, the advance was again resumed and *Tatkon* was reached. Unfortunately insufficient infantry were available to hold the town during the night, and it is beleived that 33 Jap Army HQ escaped from *Ywathit* through the town and away down to the South.

The main road South from *Tatkon* passes underneath the *Shwemyo* Bluff, which was held. This necessitated the making of a diversion over very difficult country. When it had been completed the advance continued, HQ 255 Indian Tank Brigade now being in command of the advance guard. After considerable mopping up operations North of *Pyinmana*, the column bye-passed that town and made for the *Lewe* airstip. *Toungoo* was eventually captured without opposition. There is no doubt that throughout the whole operation the speed of the advance upset all enemy calculations, and no real organized resistance on anything but a minor scale was encountered.

Meanwhile the Regiment, having completed three days much needed maintainance in their harbour in *Pyawbwe* Government Stud Farm, were under orders to move in the main body of 5 Indian Division.

On the 15th April, the Regiment on transporters less 'A' Squadron moved to M. S. 282 on the main *Rangoon* road. 'A' Squadron, moving in rear, came under the command of 123 Indian Infantry Brigade (5 Division) and halted at *Yamethin*. 'C' Squadron, after off-loading from transporters, moved to M. S. 279 & came under command of 161 Indian Infantry Brigade (5 Division). RHQ and 'B' Squadron harboured at M. S. 282.

On the 16th April 'A' Squadron moved to *Tatkon* on transporters. *RHQ*. 'B' and 'C' Squadrons did not move.

On the 17th April, Squadron *HQ*. No. 1 and No. 3 Troops of 'A' Squadron were called out to support infantry onto a hill feature North of *Kyaukse*, but were not used and the objective was reached by troops of 123 Indian Infantry Brigade with little opposition. That day Lieut M. K. Bahadur Singh M.C. left the Regiment. Shortly afterwards he succeeded his father as Maharajah of Bundi. *RHQ*. and 'B' Squadron moved to M. S. 280 a distance of only two miles. The move was necessitated by reports of bodies of tired Japs being in the area, and *HQ*. 5 Division was anxious to improve its perimeter defences. 'C' Squadron with the 4/7 Rajputs advanced to *Shwemyo*. On arrival a report was received that the *HQ*. 51 Japanese Division was situated in a nearby village. No. 1 Troop under Captain Riazul Karim Khan was sent to search for this *HQ.* but the village in which it was reported, was found to be clear. However 8 Japs were killed in a nearby nullah, one 75 mm and one 37 mm anti-tank gun were destroyed, the latter by Jemadar Mohd Ramzan. This Troop arrived in harbour after dark owing to engine trouble of two tanks.

On the 18th April, 'A' Squadron, still under the command of 123 Indian Infantry Brigade, moved to M. S. 270. On the way No. 3 Troop (Risaldar Bhag Singh) was called out to clear a large road block at M. S. 272. Infantry of 123 Brigade cleared the *Shwemyo Bluff* and directly the road was open the advance was continued behind the armoured column. 'B' Squadron moved to *Shwemyo* and came under command 9 Indian Infantry Brigade. 'C' Squadron with the 1/1 Punjab Regiment under command of 161 Indian Infantry Brigade moved along the main road to M. S. 266.

On the 19th April, RHQ moved to M. S. 272 on the *Rangoon* road and 'A' Squadron moved to M. S. 254. 'B' Squadron 'married up' at first light with the 1/1 Punjab Regiment. No. 1 Troop (Capt Baber) and No. 4 Troop (Capt Mustafa Khan) were in support of the two leading companies. The advance down to the axis of the railway line very slow because the infantry were on foot. No opposition was met until the leading troop reached the village *Pyookkwe*, which was found to be lightly held on the North side. No. 1 Troop with its supprorting infantry became involved in a small action which resulted in the killing of about 15 Japs. During the action a fire started on the back of Captain Baber's tank, but was quickly put out by Sepoy Jagir Singh, 4/4 Bombay Grenadiers, one of the escort to the tank. No. 4 Troop, which was supporting the right hand company on the West side of the railway line, was compelled to go through very thick scrub and elephant grass. Several enemy were killed. *Pyokkwe* was found to be clear, and the tanks after supporting the infantry onto a positon South of the village, where they consolidated, withdrew and harboured in the village.

'C' Squadron was split into two half squadrons. No 2 and No. 3 Troops, under command Captain Riazul Karim Khan, with 1/1 Punjab Regiment advanced across country South East of *Pyinmana.* 1700 hours this force with battery of S. P. Guns pushed accross country with the object of reaching M. S. 211. It was held up by a nullah crossing near M. S. 228 and owing to darkness harboured there. One Company 1/1 Punjab Regiment reached M. S. 228 and consolidated that position. Meanwhile No. 1 and No. 4 Troops, under command of Captain Nicolson, with one company of 4/7 Rajputs, pushed towards *Pyinmana* and cleared to within 800 yards of the town. They were engaged by one 75 mm gun firing from the South bank of the nullah, but the gun was heavily shelled and the crew were seen abondoning it. 40 Japs were estimated to have been killed.

On 20th April, RHQ moved to M. S. 249. 'A' Squadron with 123 Indian Infantry Brigade moved to *Thawatti.* 'B' Squadron moved to *Yawadaw* ahead of the infantry of 9 Indian Infantry Brigade. No opposition was met and the force harboured at *Yawadaw.* No. 1 and No. 2 Troops of 'C' Squadron acted as advance guard to 161 Indian Infantry Brigade and advanced to *Lewe* Airstrip. No. 3 and No. 4 Troops, under Captain Nicolson, supporting the 4/7 Rajputs, moved to the North of *Lewe,* coming under command of the Royal West Kents, and cleared that village Later they joined the rest of the squadron and came under command of the 1/1 Punjab Regiment.

On the 21st April, 'A' Squadron advanced to *Oktwin* and 'B' Squadron to *Lewe* Airfield with 1/1 Punjab Regiment. They rejoined RHQ in 123 Infantry Brigade harbour int he evening. Meanwhile RHQ Troop, under command of the Adjutant Capt Halliwell), was sent out to clear the diversion running (round *Pyinmana* South from the bridge at M. S. 248 and West meeting the main Road at M. S. 240. Three mines were found and lifted without incident. On arrival at *Lewe* a report was received of enemy in a village North West of the railway station. The troop with one platoon of the Royal West Kents went out to investigate. On the approach of the tanks about 50 Japs ran Westwards from the village and were shot up by the tanks, which by this time had moved to the North of the village. The troop then came back onto the main road to arbour with 161 Indian Infantry Brigade. Enemy casualties were estimated at 20.

On the 22nd April, 'A' Squadron did not move, 'B' Squadron joined up with RHQ and 'C' Squadron in the early morning. The harbour area was subjected to a Jap air attack, some casualties being caused to our troops, and one I.O.R. of H.Q. Squadron was killed. 'C' Squadron as advance guard to 161 Indian Infantry Bridge moved to *Toungoo* Airfield. RHQ and 'B' Squadron followed in the main body under command of the 1/1 Punjab Regiment

On the 23rd April, RHQ. 'B' and 'C' Squadrons remained on the *Toungoo* Airstrip. 'A' Squadron remained at *Oktwin*. 1/7 Dogra Regiment (5 Indian Division) married up with the Regiment to form part of a force known as *Smeecol* which was to push down the *Rangoon* road.

On the 24th April, *Smeecol* took over the duty of advance guard to 5 Indian Division. Its object was the capture of *Pyu*.

The force moved off in the following march:—

Advance Guard.	Composite Sqn, 7th Cav/16 Cav. 'A' Sqn. Probyn's carrying one Coy. 1/17 Dogras. Det. 36 Fd Sqn. I.E. One Bty 18th Fd Regt. R. A.
Main Body.	Scout Troop Probyn's. Coln. HQ. RHQ Troop Probyn's carrying Bn HQ 1/17 Dogras. 'C' Sqn Probyn's carrying one Coy 1/17 Dogras. 36 Fd Sqn. less Dets: 'B' Sqn Probyn's carrying one Coy 1/17 Dogras.
Rear Guard.	One Coy 1/17 Dogras. B1 Echelons of all untis.

While passing through *Toungoo*, the column was strafed by six enemy planes, which wounded one V. C. O. in 'B' Squadron and a number of I. O. Rs. of 1/17 Dogras. No opposition was met until the Pyu Chaung was reached. The road and railway bridges were found to be blown and one platoon of infantry attempted to make a bridgehead on the far side supported by tanks. This attack came under concentrated M. G. fire and the infantry had to withdraw. Artillery support and an airstrike were called for, which seemed to quieten the enemy. During the airstrike one Thunder bolt crashed on to the railway bridge. A small party of Japs, who had remained hidden on the side of the road North of the chaung, was dealt with by two sections of the tanks' escort infantry under command of the Adjutant, dismounted. All the Japs were killed. Meanwhile enemy had been located on the South side of the chaung 400 yards West of the bridge. Major Loraine-Smith while engaging the enemy was wounded by shrapnel from an enemy gun, the projectile penetrating his tank and causing a complete brew-up. The 1/17 Dogra Regiment remained on the crossing and gained a bridgehead on the other side during the night so that work on a diversion could be carried out. 'C' Squadron searched the village to the East of the road, killing Japs and taking 5 prisoners in *Kyaukkwin* village.

During the day representatives of the B. N. A. and I. N. A. were met. Capt. Britton of Anti-Fascist League Force 136, who had been dropped by parachute some three weeks previously to contact B. N. A. units, visited the harbour at M. S. 148.

On 25th April *Smeecol,* with 'B' Squadron as advance guard and *RHQ.* 'A' and 'C' Squadrons in the main body, crossed the *Pyu* Chaung a thousand yards East of the bridge and continued the advance as far as *Penwegon*, three Japs being killed en route. 1/17 Dogra Regiment occupied the village itself, the Regiment harbouring just North of the village. The task of dealing with 3,000 I. N. A. had been left to Major B. H. Mylne; the surrender is described below.

On the evening of 25th April three smartly dressed senior officers of No. 1 Mobile Indian National Army Division surrendered one mile North of the *Pyu* chaung. The names of the officers were:—

Maj. S. K. Puri-M. O. No. 2 Hospital 1 Division (formerly Capt. I. M. S. att 2/2 G. R.)

Maj. N. M. Khan O. C. No. 2 MSS 1 Div. (formerly Capt.-15 Fd. Amb.)

Capt. Gurbakhs Singh-O. C. 1 Bn. 1 Div. (formerly Jemadar-5/11 Sikh Regt.)

These officers informed Probyn's Horse that Col. Aziz Ahmed Khan, the divisional commander, was anxious to surrender with all his men, who were located in an area three miles North East of ***Pyu***. The same evening Jem. Ishar Singh of Probyn's Horse was sent to the I. N. A. HQ. with a letter instructing the divisional commander to proceed to ***Jaipur*** village the following morning at 0730 hours, under a white flag, for surrender negotiations. Major Mylne was detailed to receive the surrender. The force placed at his disposal consisted of two troops of armoured cars and one section of the Scout Troop (Lieut Bailey).

On arrival at *Jaipur*, Major Mylne selected disarming and assembly areas. The divisional commander did not arrive at the pre-arranged time so Lieut Bailey was sent to discover the reason for the delay. The divisional commander was still in bed! At 0800 hours Lieut. Bailey with Col. Aziz Ahmed Khan and his five most senior officers left for the R. V. in an ambulance. The I. N. A. officers were perfectly turned out in new uniforms and appeared to be well fed and self-confident. Major Mylne was introduced to the officers, then proceed to find out the establishment of the division, *viz*:—

H. Q. No. 1 Mobile I. N. A. Div, 3 complete guerilla regts with detachments of a fourth.

No. 1 Medical Staging Section, containing 440 patients and 1500/2000 convalescents.

Arms.	Rifles ...	.... 245.
	Bayonets ...	.... 856.
	Pistols	... 62.
	Swords ...	.... 110.
	Tommy guns ...	... 6.

Money. Two million rupees in Jap Burmese currency.

Transport.	Ambulance truck	.... 1.
	Staff cars	.... 6.
	Trucks	.... 7.
	Tonga	.... 1.
	Horses	.... 6.

Stores. Large quantities of rice, salt and dal.
Medical stores.
Ordnance stores.
Indian Independance League stores.

After the surrender had been negotiated, *HQ.* 5 Indian Division sent a party to relieve Major Mylne of the mighty task of disarming 3000 troops. It is considered that the information given by Col. Aziz Ahmed Khan was of the highest value.

On the morning of the 26th April Lieut.-Col. M. R. Smeeton, who had commanded the Regiment since August 1944, left in order to take over the temporary appointment of 2nd-in-Command of 63 Indian Infantry Brigade. *RHQ.* 'B' and 'C' Squadrons moved with 63 Indian Infantry Brigade (17 Division) to M. S. 100. 'A' Squadron moved with 48 Indian Infantry Brigade to M. S. 106.

The lead was taken over from Probyn's Horse by the Royal Deccan Horse, and the Regiment moved once again in the main body.

On 27th April *RHQ.* 'B' and 'C' Squadrons moved to M. S. 76 and 'A' Squadron to M. S. 80, moving mostly by night.

On 28th April *RHQ.* 'B' and 'C' Squadrons moved to M. S. 66½ where they were joined by 'A' Squadron.

On 29th April the Regiment again resumed the role of advance guard under command of 63 Indian Infantry Brigade and attacked *Pyagyi.* The attack, preceded by a heavy air-strike and artillery concentration was carried out frontally, 'B' Squadron on the right of the railway, 'C' Squadron between the railway and the main road. 'A' Squadron swept round the East of the village. The attack went exactly according to plan and was a perfect example of tank and infantry co-operation. The village was found to have been vacated. Either the 'preliminaries' had caused the enemy to flee or they had evacuated overnight, as the attack met no one except two dazed Burmans.

With the fall of *Payagyi* the last motorable road East

was denied to the Japs. From that time onwards country tracks were his only escape routes.

The same force continued the advance but halted South of *Payagyi* while the Regiment 'married up' with 7/10 Baluch Regiment, and a plan was made for the capture of the high ground of *Pegu*, as follows:—

Probyn's Horse less one squadron and 7/10 Baluch less one company were to move South East from the road and attack the high ground North East of *Pegu* from the East. Meanwhile 63 Indian Infantry Brigade with one squadron Probyn's Horse were to advance down the main road and join up with Probyn's and the Baluch on the objective.

The outflaking movement at first went smoothly and met no opposition. The infantry reached the objective, and a track was found by which tanks could get on to it in 'line ahead'. During consolidation heavy enemy fire opened up on to the objective from short range. It was difficult to locate it owing to thick scrub. As it was getting late the force was called back to harbour at M. S. 54. A.L.D. Gurdial Singh and Sowar Daman Singh of 'B' Sqn were wounded; three I. O. Rs. of the 4/4 Bombay Grenadier escort were killed. Meanwhile 'A' Sqn with 63 Indian Infantry Brigade reached the Northern approaches to *Pegu*, which were heavily mined. Slight artillery fire was also encountered. Much to 'A' Sqn Commander's annoyance an American newspaper reporter in his eagerness to reach *Rangoon* drove over the mines into the edge of the town. He narrowly escaped destruction by a shell and returned at great speed, again missing the mines.

On 30th April, a plan was made for the capture of *Pegu* by 17 Indian Division.

1. One Bn. of 48 Ind Inf Bde was to cross the *Pegu* River in the area of *Okpu*, advance South down the West bank and capture *Pegu* railway station.

2. 63 Ind. Inf. Bde. with one sqn Royal Deccan Horse was to capture the high ground attacked on the previous day, and exploit on down the main road as far as the road bridge over the river in the centre of the town.

3. 255 Ind. Tk. Bde. less 116 R. A. C. (Gordon

Highlanders) and one Sqn Royal Deccan Horse, with under command

18 Fd Regt. R. A.

1/3 G. R.

Composite Sqn 7 Cav/16 Cav.

were to move South East from the road at M. S. 53½ across country to attack *Kammanat*, other villages to the South, and then attack North East and meet up with 63 Ind. Inf. Bde. on the road bridge.

The move across country and capture of *Kammanat* was completed by 1130 hrs. Immediately afterwards the Regiment formed up in a line on the West edge of the village in order to attack due East into *Pegu*. The front of the objective allotted was very great so it was decided to attack only a sector of it. After considerable delay in the issuing of orders by the battalion commander an attack was put in with two squadrons and two companies up: 'A' Squadron on the left and 'C' Squadron on the right. 'B' Squadron remained in reserve. The objective was a steep ridge covered with very thick jungle. No 2 Troop (Lieut Robertson) of 'C' Squadron arrived to within 150 yards of the objective and then found the ground impassable to tanks. The squadron remained in a support role for two hours, while infantry pushed forward and attempted to observe the White Pagoda area. They suffered a number of casualties, the tanks being unable to give adequate fire support because of lack of visibility and poor communication with the infantry. 'A' Squadron succeded in getting tanks right up to the Tin Topped Pagoda area. No 1 Troop, 'B' Squadron (Capt Baber) was detached on a special mission and succeded in destroying one enemy 25 pdr gun.

In order to extend the line of the advance of the 1/3 Gurkhas, the Brigade Commander had ordered the 4/4 Bombay Granadiers less three companies to conform with the movements of the 1/3 Gurkhas and advance down the axis of the road *Thanatpin - Pegu.* It was not until 1700 hours that the 1/3 Gurkhas and the 4/4 Bombay Grenadiers were established on the objective. The tanks withdrew to harbour, as it was getting dark. The 1/3 Gurkhas sustained casualties during the evening.

Meanwhile the Royal Deccan Horse and the composite Squadron of the 7th Cavalry and the 16th Cavalry carried out their task, meeting only minor opposition, but were unable to reach the road bridge owing to a very deep nullah. They too returned to the Brigade harbour. 63 Indian Infantry Brigade reached and consolidated the area of the Golf Course but were unable to exploit further South.

The following morning patrols reported the whole of *Pegu* East of the river to be clear. All bridges had been blown and the Sappers built a Bailey bridge. On the 1st May one troop of 'B' Squadron had been held in readiness to support infantry patrols in the town but was not used.

On the 2nd May, 17 Indian Division continued its advance on *Rangoon.* 'C' Squadron with one coy 6/7 Rajputs formed part of the "advance guard mounted troops" under command of Lieut-Col. Chaudhuri O.B.E. They crossed the division made over the Pegu River at 1330 hours approximately, and advanced as far South as M. S. 39. Enemy opposition covering the road, which was extensively mined, was met at M. S. 39½. An attack was put in at 1430 hrs. No. 2 Troop (Lt. Robertson) and No 3 Troop (Ris Mohd. Arif) gave fire support to the infantry. The position was only half cleared by dark, and the force withdrew to harbour at M. S. 40½. Two 47 mm anti-tank guns and two 75mm guns were found abandoned, though they were in perfect condition. Obviously the enemy must have been in a poor state if he was unable to move or spike these guns.

The Regiment less 'C' Squadron formed part of the advance guard to 17 Indian Division under command of 48 Indian Infantry Brigade and followed *Chaudhuricol* across the river via the newly constructed Bailey bridge. Whilst combing the proposed harbour area at M. S. 47 the 6/7 Rajputs, once more co-operating with the Regiment, captured two abandoned 47mm anti-tank guns. Enemy casualties were 15 killed.

On the 3rd May, *Chaudhuricol* continued the advance, the enemy having withdrawn during the night in heavy rain. A road block was encountered at M. S. 38, and a dead Burman was found lying naked across the road. Mines were found under his body and clothing. Slight sniping was encountered, so No 4 Troop (Lt. Elder) gave fire support to the infantry while they held the side of the road, thereby allowing the sappers to clear it. The advance was very slow indeed owing to a large number of aerial bombs being dug into the road as far as *Integaw.* One troop of the 7 Cavalry and

No 1 Troop 'C' Squadron pushed through *Inegaw* without infantry and advanced to M. S. 32½, where they found that the bridge was completed demolished but not held. The remainder of the squadron advanced to the bridge and, having established a detachment of 6/7 Rajputs, withdrew to harbour at M. S. 36, where they were joined after dark by the rest of the Regiment. Owing to heavy rain it was not possible to harbour away from the main road, which was found the following morning to be crammed with transport. One troop of 36 Field Squadron had lifted over 500 mines during the day.

On the 4th May, a recce was made for a Sherman crossing of the chaung South of the bridge at M. S. 32½, but without success. Heavy rain delayed work on the crossing, so infantry was ordered to advance on foot to link up with 26 Indian Division which, having taken part in the seaborne landing near *Rangoon*, was advancing Northwards. The meeting took place at *Hlegu*. The Regiment did not move.

On the 6th May the o iginal intention that the Regiment should move to *Rangoon* was postponed due to insufficient material being available to build a Class 30 bridge. A move back to *Pegu* to join the rest of the Brigade was ordered, and sufficient accomodation was discovered there to enable the men to live under cover. The monsoon had started.

Thus ends the story of the part played by Probyn's Horse in the reconquest of Burma. In five months the Regiment moved 400 miles from its jumping-off place on the *Irrawaddy*, and a further 400 miles when fighting its way through *Meiktila* to *Pegu* in the vicinity of *Rangoon*. The Regiment arrived at *Pegu* with 98 per cent of its tanks on the road, which speaks highly for the maintenance carried out by the crews themselves, in addition to the assistance of the L. A. D. and workshops.

There is no doubt whatsoever that the employment of tanks played a major part in the destruction of the Japanese Army in *Burma*. It was fortunate that 255 Indian Tank Brigade was able to remain concentrated almost throughout and to be used in an armoured role for the capture of *Meiktila*, the prime cause of the Japanese disaster. The momentum of the final advance to *Pegu* could not have been maintained without a strong force of armour, and this advance caused the fall of *Rangoon* before the monsoon.

POST-SCRIPT

Although Burma was virtually in our hands after the fall of RANGOON, many thousands of Japs remained in the country, some of whom attempted to make their way *south east* towards MOULMEIN, which is across the SITTANG River.

On the 6th July, orders were received to send one troop of tanks to WAW, as the Japs had brought a number of guns and men across the SITTANG River in an effort to assist their men trying to escape from the PEGU YOMAS. The troop, commanded by Lt. R. Bateson, moved off on the 8th July and reached WAW late that night, coming under command of 33 Indian Infantry Brigade. This troop did not see any action as the area was entirely water-logged paddy and so they were used only as "morale boosters". The troop returned to RANGOON on the 30th July without having fired a shot.

It was expected that a large force known to be in the PEGU YOMAS would make an attempt to break out towards the SITTANG River about the 22nd July. 48 Indian Infantry Brigade was responsible for destroying any enemy who might enter their area which was both sides of the MANDALAY-RANGOON Road between DAIKU and m. s. 112. Remnants of the Jap 105 IMB, strength about 3000, were known to be in the area.

On the 16th July, 255 Indian Tank Brigade received orders to send one squadron to come under command of 48 Indian Infantry Brigade (17 Indian Division). A composite squadron was therefore formed as under, which left RANGOON with tanks on transporters on the 18th July.

Comd. Major Nicolson

Sqn HQ. 2 tanks, one each from "A" and "C" Sqns Probyn's

No. 1 Troop- "A" Sqn, Probyn's

No. 2 Troop- "C" Sqn, Probyn's

No. 3 Troop- "C" Sqn, Royal Deccan Horse

One Pl 4/4 Bombay Grenadiers

Det No. 5 Ind Mob Workshops

Det No. 1 Ind Rec Coy

On arrival the squadron was disposed as follows:—

Squadron less one troop under command 4/14 F.F.R. in PEINZALOK.

No. 3 troop (Lt. C. Peace) under command 4/17 Dogras in DAIKU.

The tanks at PEINZALOK together with a company of infantry were formed into a mobile column, with the task of dealing with any road blocks the Japs might make, or of reinforcing 48 Indian Infantry Brigade where necessary. The troop at DAIKU, where Brigade HQ. was located, was a reserve for any contingency for the Southern of the brigade area.

At 0500 hours on the 22nd July, enemy were seen crossing the railway line from West to East by the bridge at m.s. 102½. No. 1 Troop (Capt. Kaushal) and the Squadron HQ. tanks moved down the line from their harbour in the railway station and engaged about 100 enemy retreating North-East at ranges from 600 to 2000 yards. Shooting was obstructed by the bushes on the banks of the OULMETHA Chaung and fire positions, were difficult to find. The Squadron Commander's tank became bogged and control then operated from a jeep. After all enemy movement had ceased, a section of the Bombay Grenadiers was sent along the banks of the choung West of the railway bridge. They killed 12 Japs and captured six more, including one officer. A further patrol moved to within 600 yards East of the railway and counted 16 dead. It captured two wounded prisoners. The next day another patrol counted 25 more dead further out in the paddy, and it was evident that villagers had thrown dead Jap bodies into the PEINZALOK Chaung.

At 0600 hours on the 22nd July, while the action described in the previous paragraph was taking place, information was received that the Japs had approached the road at M. S. 114. Consequently No. 2 Troop (Lieut Fitzherbert) was ordered out to assist a platoon of the 7/10 Baluch who were in this area. He was later joined by Lieut. Swan in a Sqn H. Q. tank. The Japs had reached GULAB and RAM DEO villages in the night, and were caught trying to cross the road at first light. The tanks opened fire from the road at M. S. 114 at the enemy in and around RAM DEO village and also at enemy retreating across the baddy from GULAB to SHEO GHULAM. Excellent shoots were obtained and all enemy seen were liquidated.

A patrol of the 7/10 Baluch suffered casualties when trying to clear the villages and withdrew. Pin point targets as indicated but the patrol were then engaged. Unfortunately the state of the ground did not permit the tanks to enter the village in support of the infantry, and at 1130 hours the tanks were withdrawn. When the village was occupied in the evening over 70 enemy dead were counted inside and around it.

No further opportunities for the employment of tanks were presented, and the whole squadron returned to RANGOON on the 23rd July.

APPENDIX A.

Officers of Probyn's Horse who participated in the crossing of the Irrawaddy.

Lt. Col. M. R. Smeeton, MBE,MC	....	Commandant
Major F. W. Kennedy ...	....	Second-in-Comd.
Major B. H. Mylne MBE ...	...	'HQ' Sqn. Comdr.
Major B. L. Loraine-Smith	....	'A' Sqn. Comdr.
Major W. M. Arkinstall ...	....	'C' Sqn. Comdr.
Major H.I.E.R.C. Stewart	....	'B' Sqn. Comdr.
Capt. Udham Singh ...	....	'B' Sqn. 2 I/C.
Capt. Bhagat Singh	...	'HQ' Sqn. 2 I/C.
Capt. M. A. Nicolson ...	...	'A' Sqn. 2 I/C.
Capt. A. D. O'Beirne-Ryan	...	'C' Sqn. 2 I/C.
Capt. Ihsan Ullah Baber ...	...	'B' Sqn. Tp. Ldr.
Capt. Riaz-ul Karim Khan ...	....	'C' Sqn. Tp. Ldr.
Capt. P. S. Fallows ...	....	Quartermaster.
Capt. F. P. Sedman	...	Signal Officer.
Capt. R. D. Anderson ...	....	'C' Sqn. Tp. Ldr.
Capt. E. Halliwell ...	...	Adjutant.
Capt. Pritpal Singh ...	....	Technical Officer.
Lieut J. de B. Carey	....	'B' Sqn. Tp. Ldr.
Lieut A. W. Lane	...	'A' Sqn. Tp. Ldr.
Lieut F. G. Turner ...	...	'B' Sqn. Tech. Officer.
Lieut J. A. Chiles	....	'C' Sqn. Tech. Officer.
Lieut K. K. Grover	....	'B' Sqn. Tp. Ldr.
Lieut F. R. Wish ...	....	'A' Sqn. Tech. Officer
Lieut S. E. Parkes	....	'HQ' Sqn. Tech. Offr.
Lieut M. K. Bahadur Singh	....	'A' Sqn. Tp. Ldr.
Lieut W. C. Keown	....	'A' Sqn Reinforcement.
Lieut D. A. Edwards	....	'Tp. Ldr. 'C' Sqn.
Lieut C. J. Bailey ...	...	Scout Tp. O. C.
Lieut G. L. Quinn ...	....	Scout Tp. 2 I/C.
Lieut K. Swindells	....	L.O. 255 Ind. Tk. Bde.
Lieut R. Bateson	....	'A' Sqn Reinforcement.
Lieut J. A. Elder	....	'C' Sqn. Tp. Ldr.
Lieut G. Gilkison	....	Intelligence Officer.

2/Lieut N. M. Robertson	....	Tp. Ldr. 255 Ind. Tk. Bde.
2/Lieut J. R. Finlow	....	'HQ' Sqn. Echlon Offr.
Capt. D. V. Bapat IAMC	....	Regtl. M. O.
Capt. J. Coyne REME	....	O. C. L. A. D.

Changes in Officers up to Leaving Meiktila.

Capt. Riazul Karim Khan 'C' Sqn. 2-I/C vice Capt. A. D. O'Beirne-Ryan killed in action.

Lieut W. C. Keown proceeded on F. V. S. Course (Ahmednagar) and subsequently rejoined at *Pegu* 11 May 45.

Capt. R. D. Anderson killed in action.

Lieut J. de B. Carey wounded and evacuated

Lieut D. A. Edwards wounded and evacuated

Lieut K. Swindells killed in action.

Lieut R. Bateson wounded and evacuated whilst with 116 R. A. C. at *Taungtha*.

Lieut G. Gilkison became spare Troop Leader in 'A' Sqn.

Lieut J. R. Finlow became Intelligence Officer vice Lieut G. Gilkison.

2/Lieut D. J. Macgill joined Regiment on the 12th March and went to 255 Ind. Tank Bde. as L.O. as relief for Lieut K. Swindells who was killed in action on being transferred from Bde. H. Q. to 'A' Sqn.

Capt. Mustafa Khan joined Regt. from 43 Cavalry on the 2nd April and became Tp. Leader in 'B' Sqn.

Major F W. Kennedy proceeded on 21 days War Leave on the 3rd April.

Officers of Probyn's Horse who were with the Regt. on leaving Meiktila.

Lt.-Col. M. R. Smeeton, DSO,MBE,MC		Commandant
Major B.H. Mylne, MBE	....	'HQ' Sqn Comdr.
Major B. L. Loraine-Smith, MC	....	'A' Sqn Comdr.
Major W.M. Arkinstall	....	'C' Sqn Comdr.
Major H.I.E.R.C. Stewart, MC	....	'B' Sqn Comdr.
Capt Mustafa Khan	....	'B' Sqn Tp. Ldr.
Capt Udham Singh	....	'B' Sqn 2 I/C

Capt Bhagat Singh	.... 'HQ' Sqn 2 I/C
Capt M.A. Nicolson	... 'A' Sqn 2 I/C
Capt Ihsan Ullah Baber	.... 'B' Sqn Tp. Ldr.
Capt Riazul Karim Khan, MC	... 'C' Sqn 2 I/C
Capt P.S. Fallows ...	... Quartermaster
Capt F.P. Sedman	.... Signal Officer
Capt E. Halliwell	... Adjutant
Capt A.W. Lane, MC ..	.... Tp. Ldr.
Capt Pritpal Singh ...	.... Technical Officer
Lieut F.G. Turner ...	.... 'B' Sqn Tech. Offr
Lieut J.A. Chiles ...	... 'C' Sqn Tech. Offr
Lieut F R. Wish	.... 'A' Sqn Tech. Offr
Lieut M.K. Bahadur Singh, MC	.... 'A' Sqn Tp. Ldr
Lieut S.E. Parkes	.... 'HQ' Sqn Tech, Offr
Lieut C.J. Bailey ...	.... Scout Tp. O.C.
Lieut G.L. Quinn	.... Scout Tp 2 I/C
Lieut J.A. Elder	.... 'C' Sqn Tp. Ldr
Lieut G. Gilkison ...	.... 'A' Sqn Tp. Ldr (Spare)
Lieut N.M. Robertson	.... 'C' Sqn Tp. Ldr
Lieut J.R. Finlow	.... Intelligence Officer
Lieut D.J. MacGill	.... L.O. 255 Ind. Tk. Bd.
Capt D.V. Bapat IAMC	... Regtl MO.
Capt J. Coyne REME	... O.C. L.A.D.

Changes in Officers up to arrival Pegu 6 May, 1945.

Lt.Col M.R. Smeeton, DSO, MBE, MC was posted as Second-in-Command 63 *Indian Infantry Brigade* (17 Ind Division) on 28.4.45; Major F. W. Kennedy became officiating Commandant.

Major F. W. Kennedy rejoined the Regt from War Leave on 28. 4. 45.

Major W. M. Arkinstall killed in action by enemy mine 11. 4. 45.

Capt F. P. Sedman evacuated Sick on 9.4.45.

Lieut K. K. Grover killed in action on 10.4.45.

Lieut B. I. Fitzherbert joined the Regt on 6.4.45.

2/Lieuts Amarjit Singh and Ashok Raje Dhairyashilrao Gaekwar joined the Regt on 10.4.45.

Lieut M. K. Bahadur Singh, MC proceeded on indefinate leave 18.4.45.

Major B. L. Loraine-Smith, MC. wounded in action and evacuated on 24.4.45.

APPENDIX B.

OFFICERS KILLED IN ACTION.

E.C.1331	Lieut. K. Swindells	17-3-45
E.C.182997	Capt. D. O'Beirne-Ryan	23-3-45
I.E.C. 2569	Lieut. K. K. Grover	10-4-45
E.C.170929	Major W.M. Arkinstall	11-4-45

OFFICERS DIED OF WOUNDS

		Wounded	Died
E.C. 4220	Capt. R. D. Anderson	28-2-45	12-3-45

OFFICERS WOUNDED AND EVACUATED.

E.C 12744	Lieut. D. A. Edwards	14-3-45
E C.4663	Lieut. J. de B. Carey	22-3-45
E.C.14994	Lieut. R. Bateson	28-3-45
575 AI	Major B. L. Loraine-Smith	24-4-45

OFFICERS WOUNDED BUT NOT EVACUATED

Lt. Col.	M. R. Smeeton
Lieut	J. A. Elder
Lieut	Bahadur Singh
Capt.	Udham Singh

INDIAN RANKS KILLED IN ACTION

3124	Dfr.	Durga Singh	22-2-45
15648	Swr.	Gurbachan Singh	10-3-45
16001	Swr.	Mubarak Ali	23-3-45
47735	Swr.	Ajit Singh	25-3-45
10101	ALD.	Hadayat Ahmed	8-4-45
16072	Swr.	Gajraj Singh	22-4-45

INDIAN RANKS DIED OF WOUNDS

			Wounded	Died
14495	Swr.	Dayal Singh	22-2-45	23-2-45
48507	ALD.	Fazal Beg	23-3-45	23-3-45
40462	Swr.	Onkar Singh	8-4-45	9-4-45
4158	ALD.	Harcharan Singh	9-4-45	16-4-45
48493	Swr.	Abdul Rehman	11-4-45	12-4-45

INDIAN RANKS WOUNDED AND EVACUATED.

2918	L. Dfr	Risal Singh	1-3-45
5433	L. Dfr	Mohd Ashraf	3-3-45
	Jem.	Kehar Singh	11-3-45
	Jem.	Mehar Singh	10-3-45
47015	Swr.	Karnail Singh	10-3-45
14898	Swr.	Daya Singh	11-3-45
	Ris.	Shamsher Singh (B Sqn.)	11-3-45
15565	Swr.	Shah Hussain	14-3-45
17368	Dfr.	Mohd. Yasin	18-3-45
3715	L. Dfr	Bostan Khan	18-3-45
15687	Swr.	Mohd Zaman	18-3-45
3978	L. Dfr	Sardar Khan	23-3-45
6674	Swr.	Dost Mohd	23-3-45
13597	Swr.	Mohd. Illahi	23-3-45
16039	Swr.	Dhakan Singh	23-3-45
3267	Dfr.	Adalat Khan	24-3-45
7967	Swr.	Mohd. Sheer	24-3-45
2850	ALD.	Nazar Singh	25-3-45
17780	Swr.	Ujagar Singh	27-3-45
9641	ALD.	Midda Singh	27-3-45
	Jem.	Gurbachan Singh	24-4-45
20377	Swr.	Nasir Ahmed	9-4-45
	Jem.	Bhag Singh	9-4-45
	Jem.	Sikandar Singh	9-4-45
	Ris.	Ujagar Singh	10-4-45
10491	Swr.	Rumal Singh	23-3-45
9692	Swr.	Daman Singh	29-4-45
9712	Swr.	Gurdial Singh	29-4-45
4497	ALD.	Mela Ram (not evacuated)	1-5-45

APPENDIX 'C'

DECORATIONS.

The following immediate awards have been granted to personnel of Probyn's Horse during the Burma Campaign.

D.S.O. Lieut-Colonel M.R. Smeeton MBE, MC.

M.C. Major B.L. Loraine-Smith.
Major H.I.E.R.C. Stewart.
Capt. Riazul Karim Khan.
Lieut A. W. Lane.
Lieut Bahadur Singh
Ris Siri Singh
Jem Gurbachan Singh

M.M. L/Dfr Ram Lall (4281)
A.L. Dfr Painda Khan (15564)
L/Dfr Baboo Singh (1470)

MENTION IN DESPATCHES.

Lieut R. Bateson.
Jem Bhag Singh.

CERTIFICATES of GALLANTRY.

A.L. Dfr Sita Ram (4399)
A.L. Dfr Kishan Singh (A12137)

Note :—The results of recommendations for periodical awards have so far not been received.

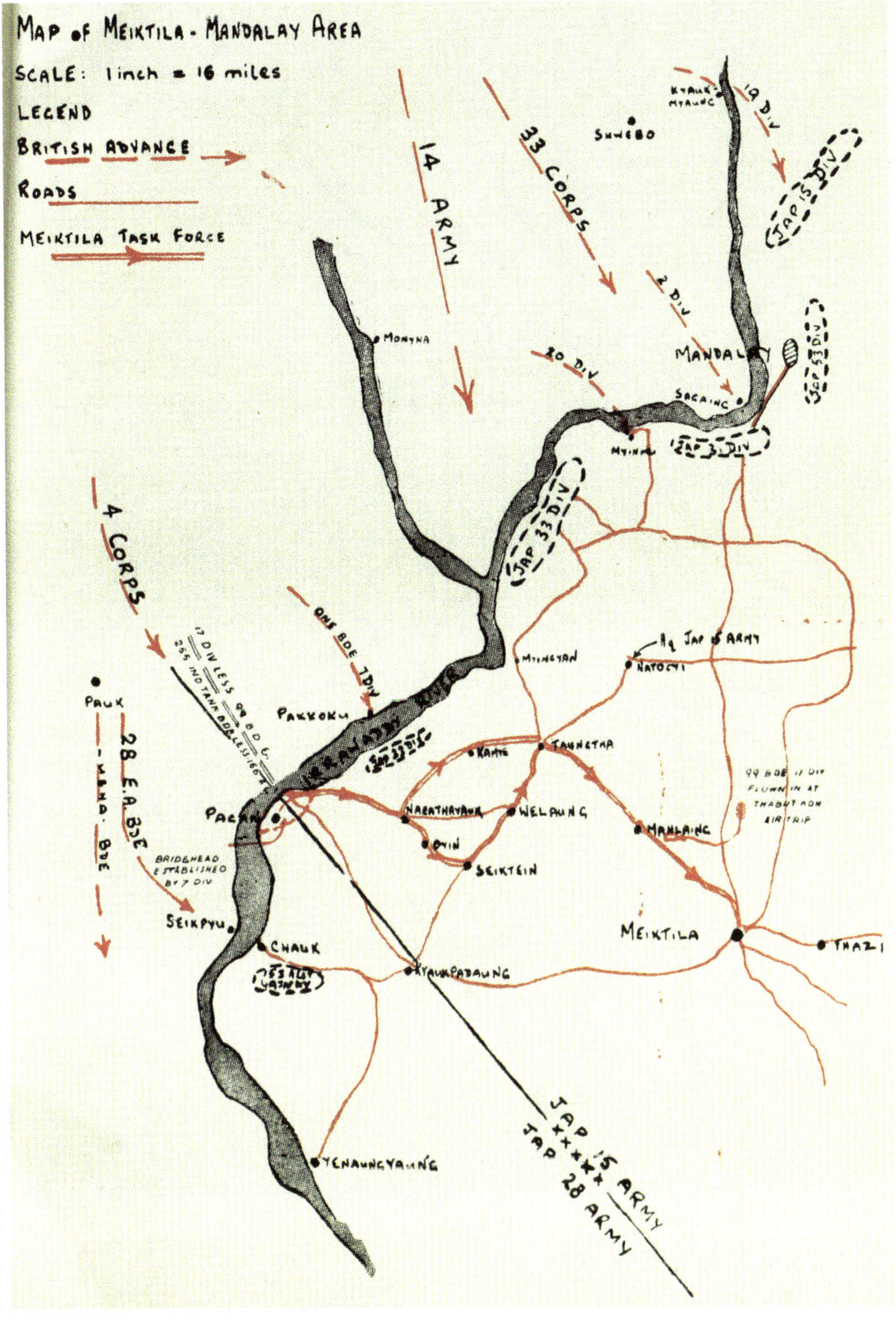

MAP OF MEIKTILA - MANDALAY AREA

SCALE: 1 inch = 16 miles

LEGEND

BRITISH ADVANCE

ROADS

MEIKTILA TASK FORCE

14 ARMY

33 CORPS

KYAUK-MYAUNG

19 DIV

SHWEBO

JAP 15 DIV

2 DIV

MONYWA

20 DIV

MANDALAY

JAP 53 DIV

SAGAING

MYINMU

JAP 31 DIV

JAP 33 DIV

4 CORPS

ONE BDE 7 DIV

17 DIV LESS 99 BDE & 255 IND TANK BDE LESS 116 R

MYINGYAN

Hq JAP 15 ARMY

NATOGYI

PAUK

PAKKOKU

IRRAWADDY RIVER

JAP 33 DIV

KAMYE

TAUNGTHA

28 E.A. BDE

LUSHAI BDE

99 BDE 17 DIV FLOWN IN AT THABUTKON AIRSTRIP

PAGAN

NYAUNGU

WELAUNG

MAHLAING

BRIDGEHEAD ESTABLISHED BY 7 DIV

SEIKTEIN

MEIKTILA

SEIKPYU

CHAUK

THAZI

53 ALG MOBY

KYAUKPADAUNG

YENANGYAUNG

JAP 15 ARMY

JAP 28 ARMY

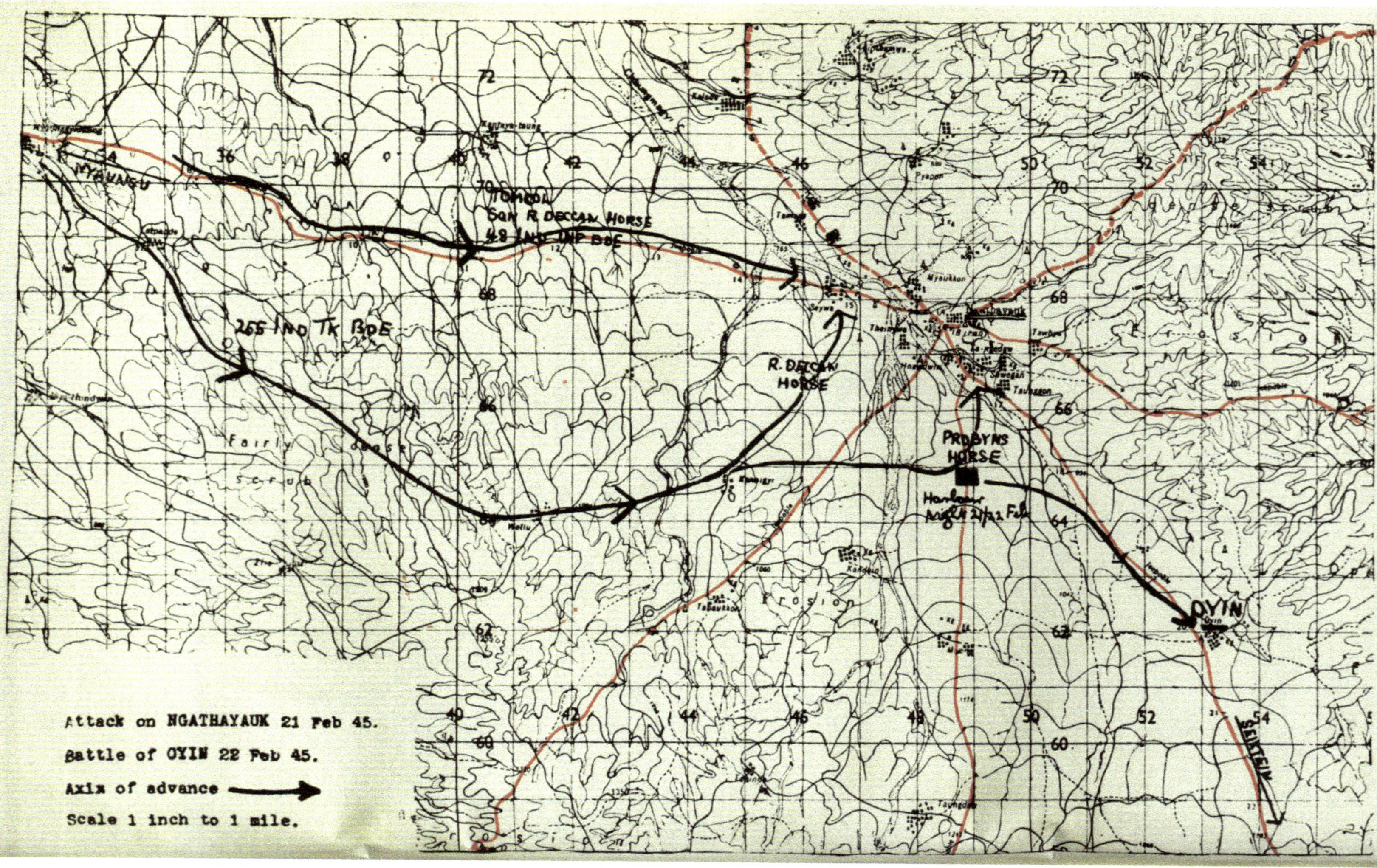
MYAUNGU
TOMCOL
SQN R DECCAN HORSE
48 IND INF BDE
255 IND TK BDE
R. DECCAN HORSE
PROBYNS HORSE
Harbour night 21/22 Feb
OYIN
SEIKTEIN
Fairly dense scrub
Attack on NGATHAYAUK 21 Feb 45.
Battle of OYIN 22 Feb 45.
Axis of advance
Scale 1 inch to 1 mile.

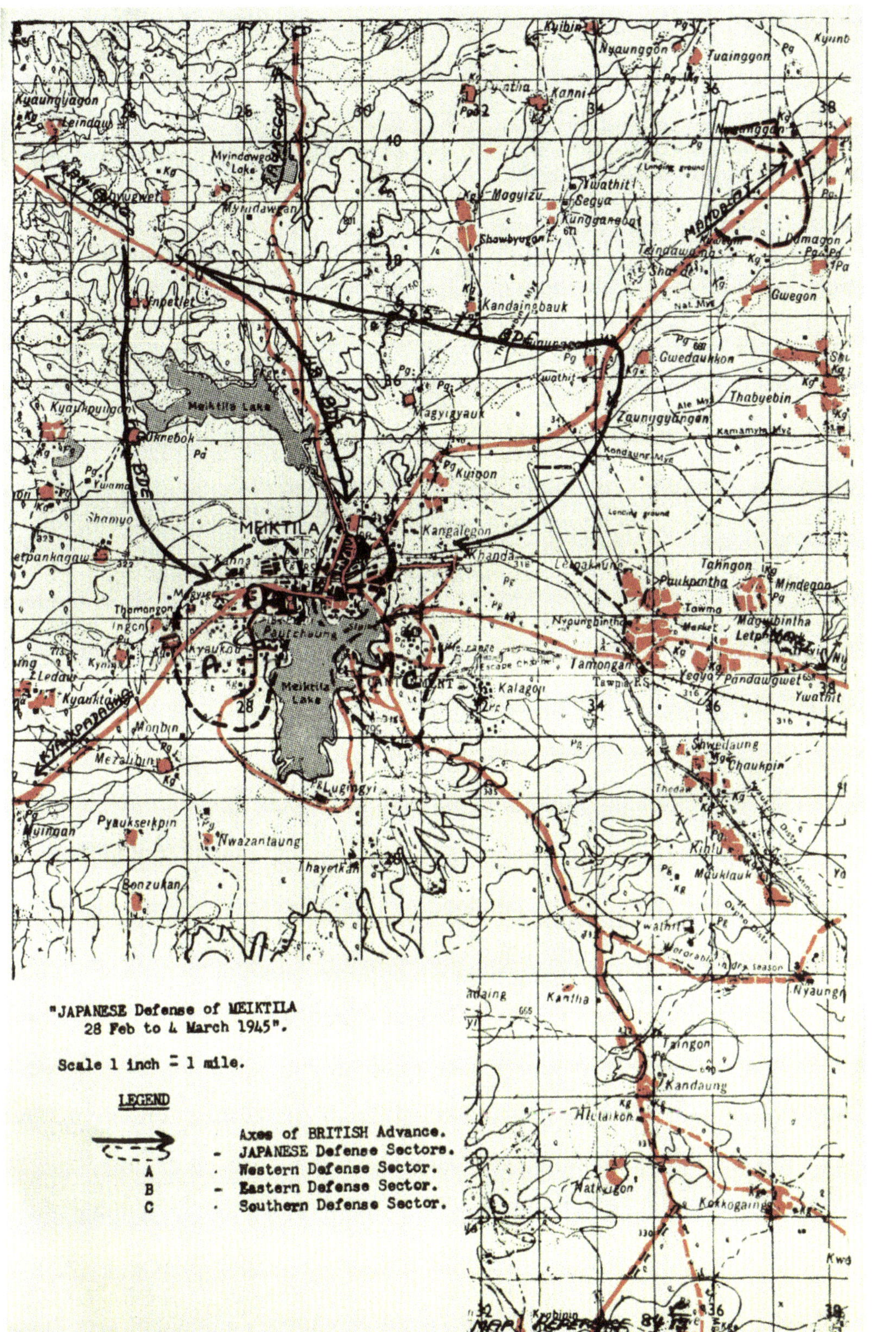
"JAPANESE Defense of MEIKTILA
28 Feb to 4 March 1945".
Scale 1 inch = 1 mile.
LEGEND
Axes of BRITISH Advance.
- JAPANESE Defense Sectors.
A . Western Defense Sector.
B - Eastern Defense Sector.
C . Southern Defense Sector.
MEIKTILA
Meiktila Lake
Meiktila Lake
CANTONMENT
MANDALAY
KYAUKPADAUNG
Kyaungyagon
Leindaw
Myindawgan Lake
Myindawgan
Kyibin
Nyaunggon
Tuainggon
Kanni
Magyizu
Segya
Kunggangon
Shawbyugon
Kandaingbauk
Kyaukpyugon
Shamyo
Kyigon
Kangalegon
Khanda
Zaunggyangon
Thabyebin
Gwedaukkon
Gwegon
Taingon
Paukpintha
Mindegon
Magyibintha
Tamongan
Yegya
Pandawgwet
Ywathit
Kalagon
Paukchaung
Kyaukon
Thamangon
Monbin
Mezalibin
Kyauktaung
Lugingyi
Pyaukseikpin
Nwazantaung
Thayetkan
Benzukan
Shwedaung
Chaukpin
Kihlu
Mauklauk
Nyaungon
Kantha
Taingon
Kandaung
Htataikon
Natkyigon
Kokkogaing
MAP REFERENCE 84 F5

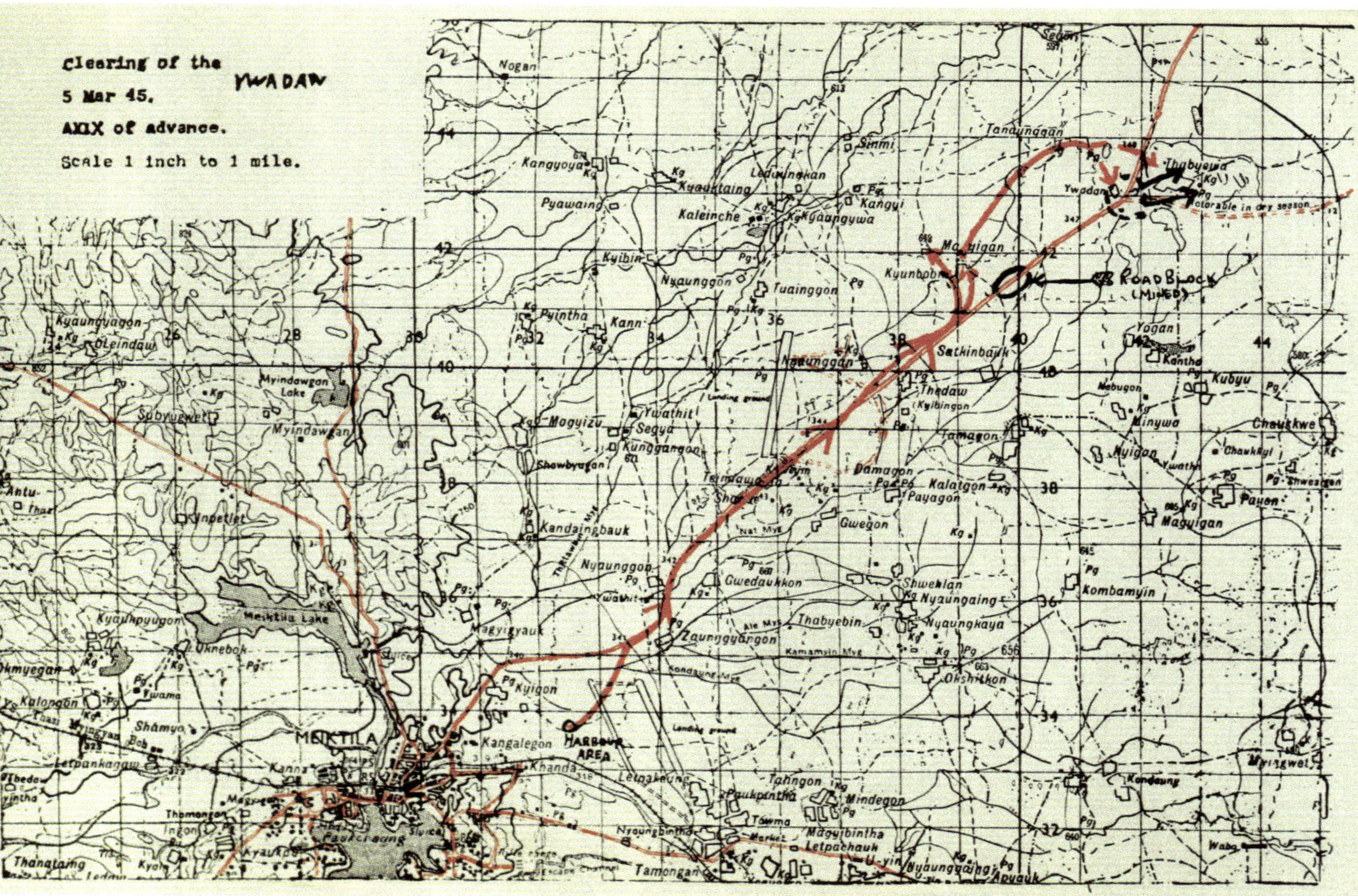

Clearing of the YWADAW
5 Mar 45.
AXIX of advance.
Scale 1 inch to 1 mile.

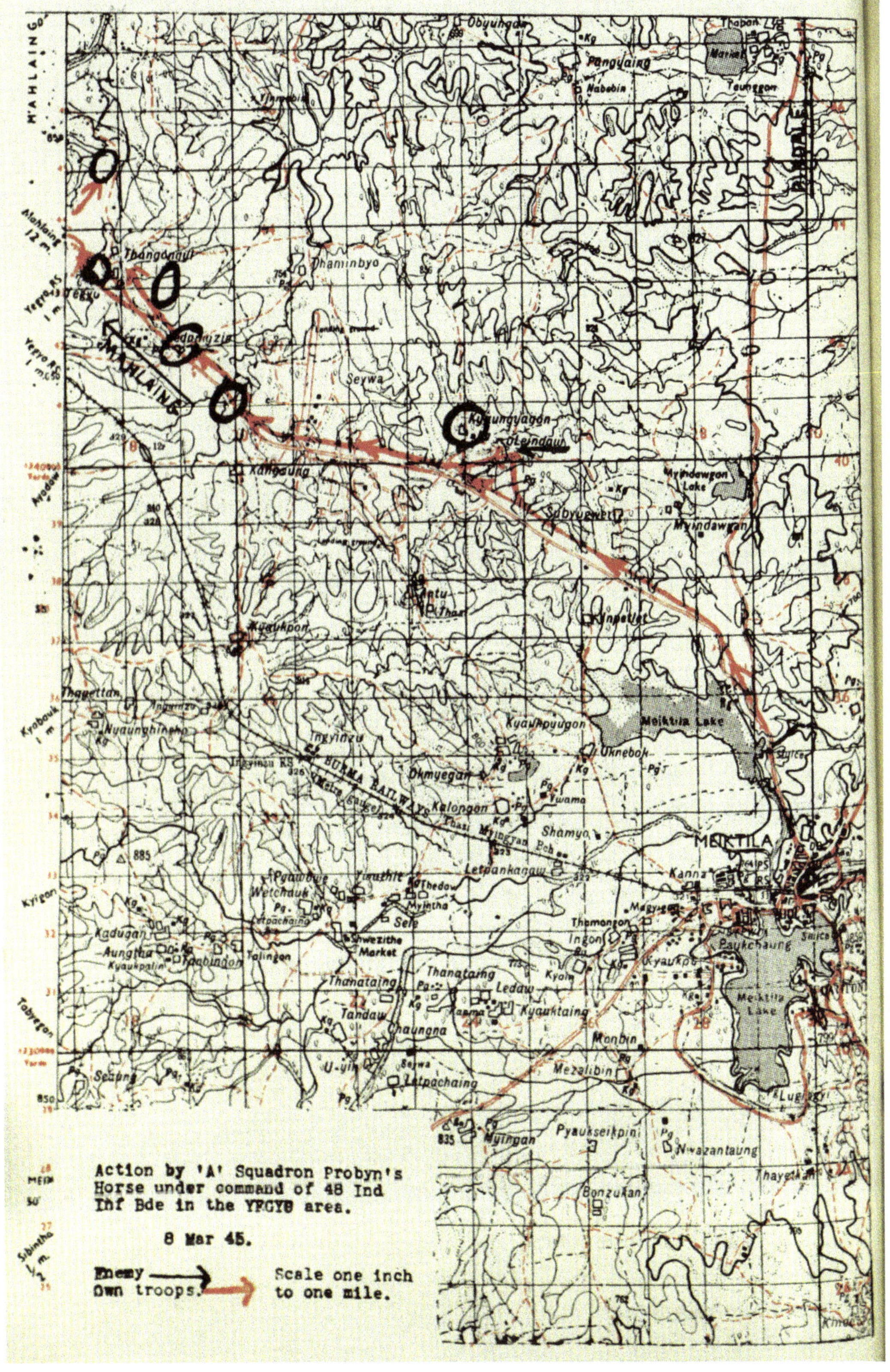
Obyuhnga
Pangulaing
Nabebin
Thapan
Taunggon
Thangongut
Dhaminbyo
Seywa
Kyaungyaggn
Leindaw
Kangaung
Sabyugwet
Myindawgon Lake
Myindawgan
Antu
Kinpetlet
Kyaukpon
Thagettan
Nyaunghinpho
Ingyinzu
Kyaukpyugon
Meiktila Lake
Ukneboh
Okmyegan
BURMA RAILWAYS
Kalongon
Shamyo
MEIKTILA
Letpankangaw
Thedaw
Myintha
Sele
Wetchauk
Kadugan
Aungtha
Shwezithe Market
Thanataing
Ledaw
Kyauktaing
Tandaw
Chaungna
Monbin
Mezalibin
Letpachaing
Myingan
Pyaukseikpin
Nwazantaung
Thayettan
Bonzukan
MAHLAING
Action by 'A' Squadron Probyn's Horse under command of 48 Ind Inf Bde in the YEGYO area.
8 Mar 45.
Enemy
Own troops.
Scale one inch to one mile.

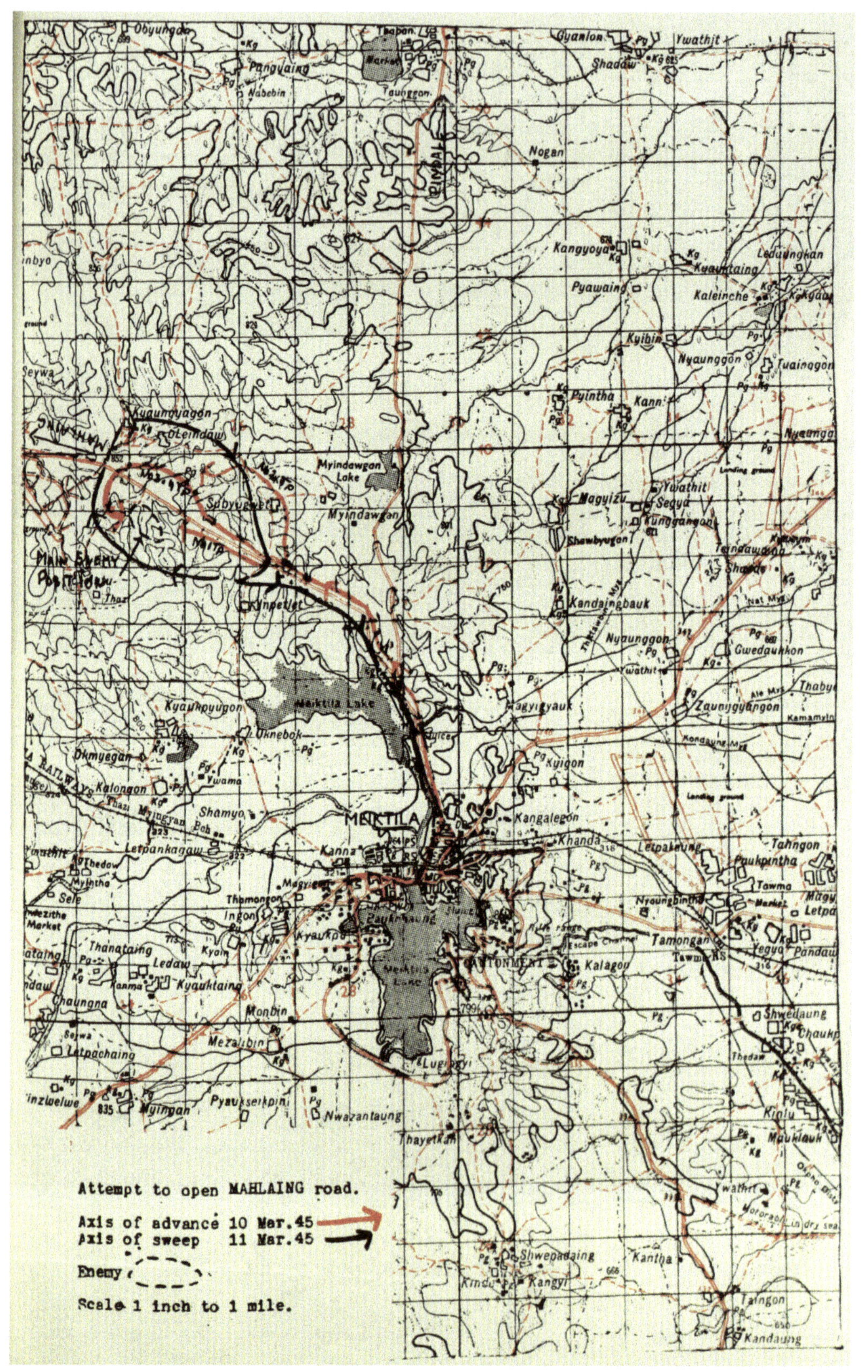
Obyunqon
Thabon
Gyanlon
Ywathit
Shadaw
Pangyaing
Nabebin
Market
Taunggon
PINDALE
Nogan
Kangyoya
Ledaungkan
Kyauktaing
Pyawaing
Kaleinche
Kyibin
Nyaunggon
Tuainogon
Pyintha
Kann
Kyaungyagon
Leindaw
Myindawgon Lake
Myindawgon
Sabyugwet
Magyizu
Ywathit
Segya
Kunggangon
Shawbyugon
Main Enemy Posn
Kinpetlet
Kandaingbauk
Nyaunggon
Gwedaukkon
Kyaukpyugon
Meiktila Lake
Ukneboh
Magyigyauk
Zaunggyangon
Okmyegan
Kyigon
Kalongon
Ywama
Shamyo
Landing ground
MEIKTILA
Kangalegon
Kanna
Khanda
Letpankagaw
Letpakaung
Tahngon
Paukpintha
Thedow
Mylntha
Sele
Thamongon
Ingon
Nyaungbintha
Magyigon
Paukkaung
Kyaukpu
Tamongan
Thanataing
Kyain
Ledaw
CANTONMENT
Kalagon
Kyauktaing
Chaungna
Monbin
Shwedaung
Mezalibin
Letpachaing
Lugyi
Myingan
Pyaukserkpin
Nwazantaung
Kinlu
Thayetkan
Mauklauk
Attempt to open MAHLAING road.
Axis of advance 10 Mar.45
Axis of sweep 11 Mar.45
Enemy
Scale 1 inch to 1 mile.
Shwepadaing
Kantha
Kindu
Kangyi
Taingon
Kandaung

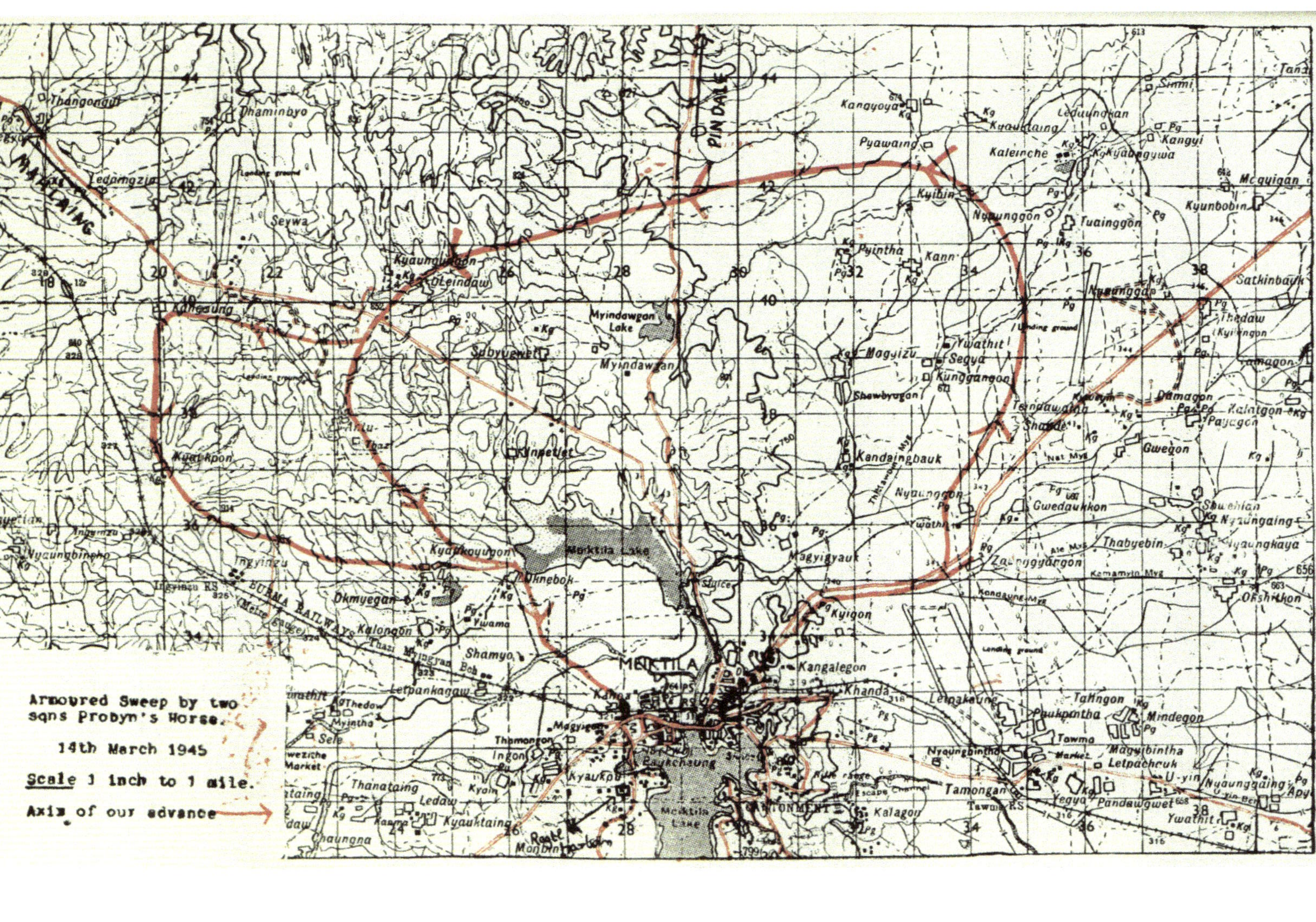
Armoured Sweep by two sqns Probyn's Horse.
14th March 1945
Scale 1 inch to 1 mile.
Axis of our advance
MEIKTILA
PINDALE
MAHLAING
Meiktila Lake
Myindawgan Lake
BURMA RAILWAYS (Metre Gauge)
CANTONMENT
Thangongon
Dhaminbyo
Seywa
Kyaungon
Leindaw
Subyuwet
Myindawgan
Kinpetlet
Kyaukkoyugon
Oknebok
Dkmyegan
Kalongon
Shamyo
Letpankagaw
Kyaukpon
Thanataing
Ledaw
Kyauktaing
Kanayoya
Pyawaing
Kyibin
Nyaunggon
Tuainggon
Kaleinche
Kyunbobin
Mcquigan
Pyintha
Kann
Magyizu
Ywathit
Segya
Kunggangon
Shawbyugon
Kandaingbauk
Magyigyauk
Kyigon
Kangalegon
Khanda
Satkinbauk
Thedaw
Damagon
Payegon
Gwegon
Gwedaukkon
Thabyebin
Zaungyargon
Ofshithon
Tahngon
Mindegon
Pauknintha
Towma
Magyibintha
Letpachauk
Tamongan
Kalagon
Yegya
Pandawgwet
Ywathit
Nyaunggaing

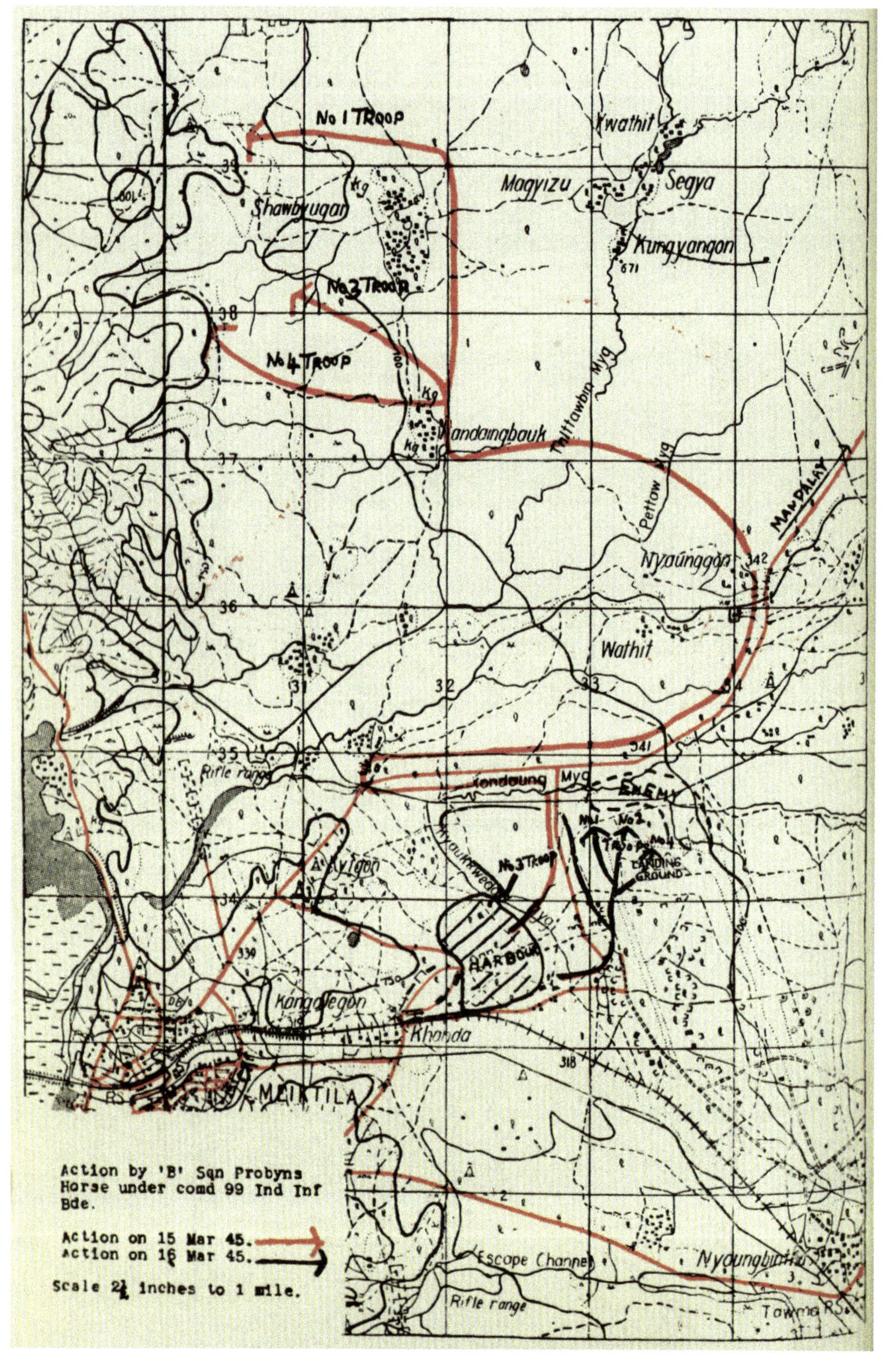

No 1 TROOP
No 3 TROOP
No 4 TROOP
No 3 TROOP
Shawbyugan
Kwathit
Magyizu
Segya
Kungyangon
Kandaingbauk
Thittawbin Myg
Pettow Myg
MANDALAY
Nyaunggon
Wathit
Rifle range
Kondaung Myg
ENEMY
LANDING GROUND
HARBOUR
Kangyegon
Khanda
MEIKTILA
Escape Channel
Nyaungbintha
Rifle range
Tawma RS
Action by 'B' Sqn Probyns Horse under comd 99 Ind Inf Bde.
Action on 15 Mar 45.
Action on 16 Mar 45.
Scale 2½ inches to 1 mile.

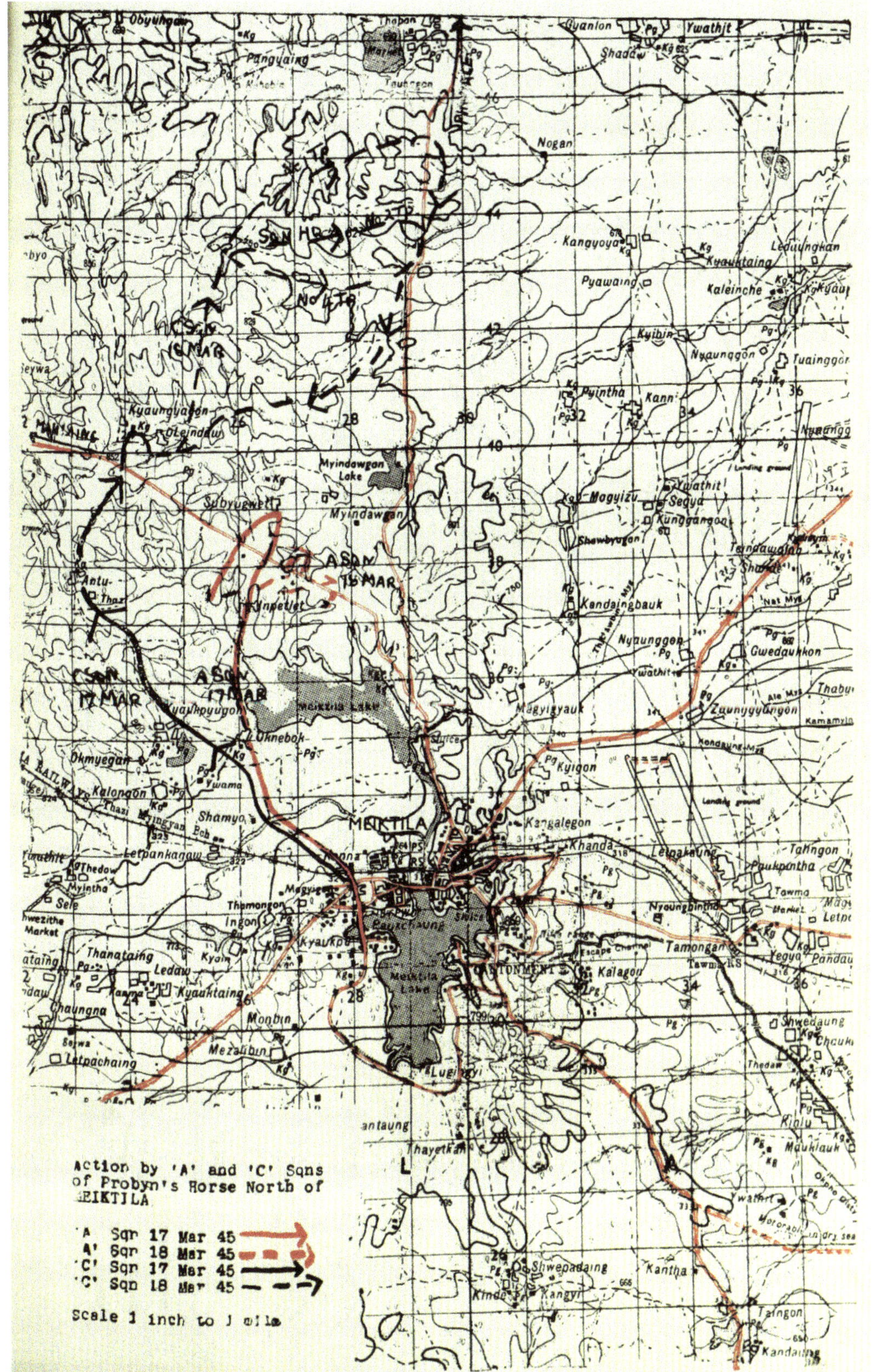
Action by 'A' and 'C' Sqns
of Probyn's Horse North of
EIKTILA
'A' Sqn 17 Mar 45
'A' Sqn 18 Mar 45
'C' Sqn 17 Mar 45
'C' Sqn 18 Mar 45
Scale 1 inch to 1 mile
MEIKTILA
Meiktila Lake
CANTONMENT
Myindawgan Lake
A SQN 18 MAR
A SQN 17 MAR
C SQN 17 MAR
SQN HQ
Kyaukpyugon
Ukneboh
Kyigon
Kangalegon
Khanda
Letpankagaw
Thamongon
Ingon
Kyaukpo
Monbin
Mezalibin
Lugiskyi
Thayetkan
Shwepadaing
Kantha
Kangyi
Taingon
Kandaung
Nyaunggon
Gwedaukkon
Zaungyyangon
Magyiyauk
Kandaingbauk
Shawbyugon
Magyizu
Segyu
Kunggangon
Kangyoya
Pyawaing
Kyibin
Pyintha
Kann
Nogan
Shaddw
Ywathit
Kaleinche
Kyauktaing
Leduungkan
Tuainggon
Pangyuaing
Kyaungyaogon
Leindaw
Antu
Subyugwe
Kinpetlet
Okmyegan
Kalongon
Ywama
Shamyo
Thedow
Myintha
Sele
Thanataing
Ledaw
Kyauktaing
Letpachaing
Chaungna
Kalagon
Tamongan
Nyaungbintha
Letpakaung
Paukpintha
Taingon
Shwedaung
Kinlu
Mauklaug
Ywathit

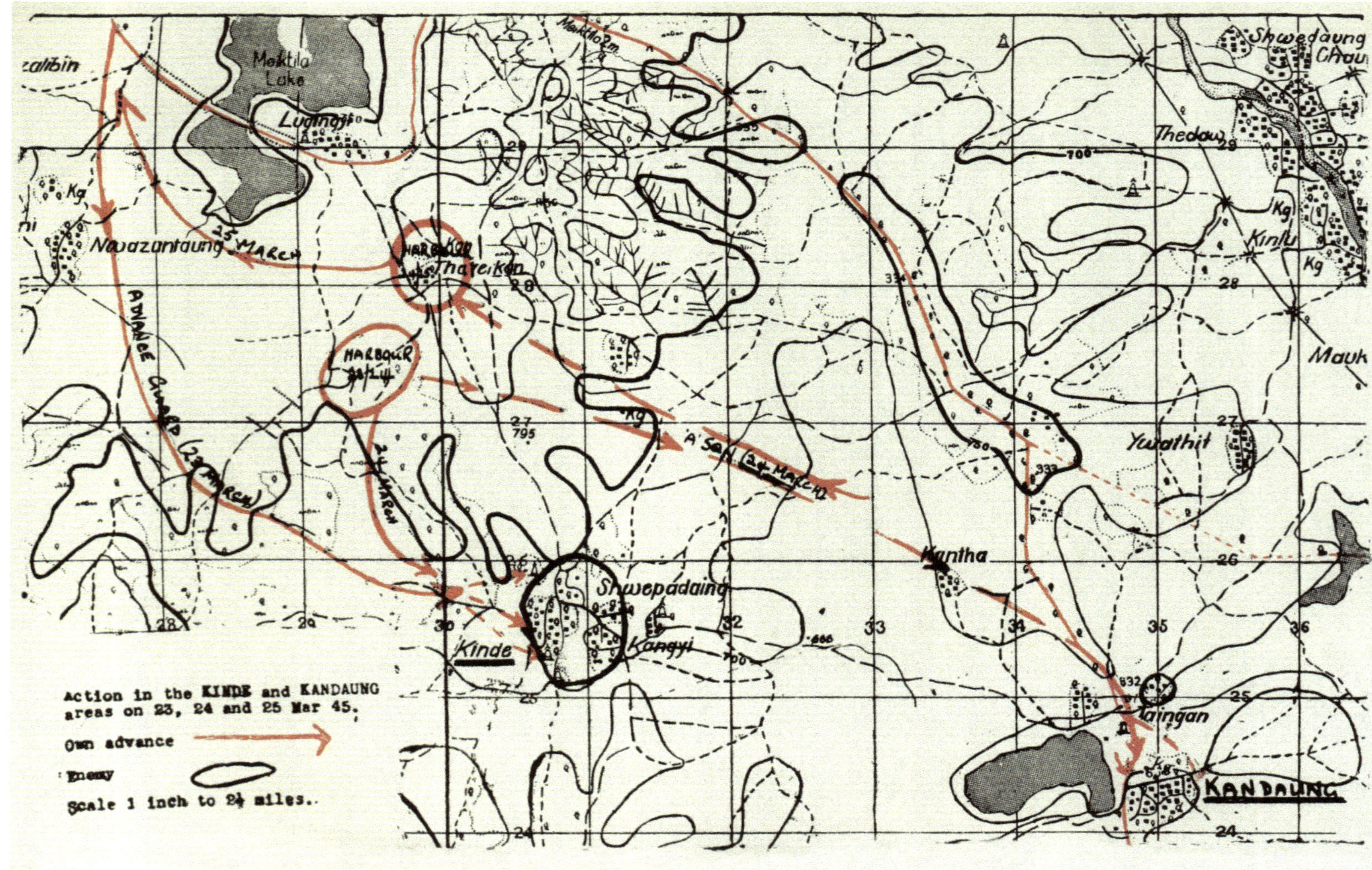
Meiktila Lake
Luginagyi
Nawazuntaung
25 MARCH
ADVANCE GUARD (23 MARCH)
HARBOUR
Thareikon
HARBOUR 23/24
24 MARCH
A'SAN (24 MARCH)
Kinde
Shwepadaing
Kangyi
Kantha
Ywathit
Thedaw
Kinlu
Mauk
Shwedaung
Laingan
KANDAUNG
Action in the KINDE and KANDAUNG
areas on 23, 24 and 25 Mar 45.
Own advance
Enemy
Scale 1 inch to 2¼ miles.

MAP SHOWING THE 'LEFT HOOK' UNDERTAKEN
BY CLAUDCOL PRIOR TO THE CAPTURE OF PYAWBWE.

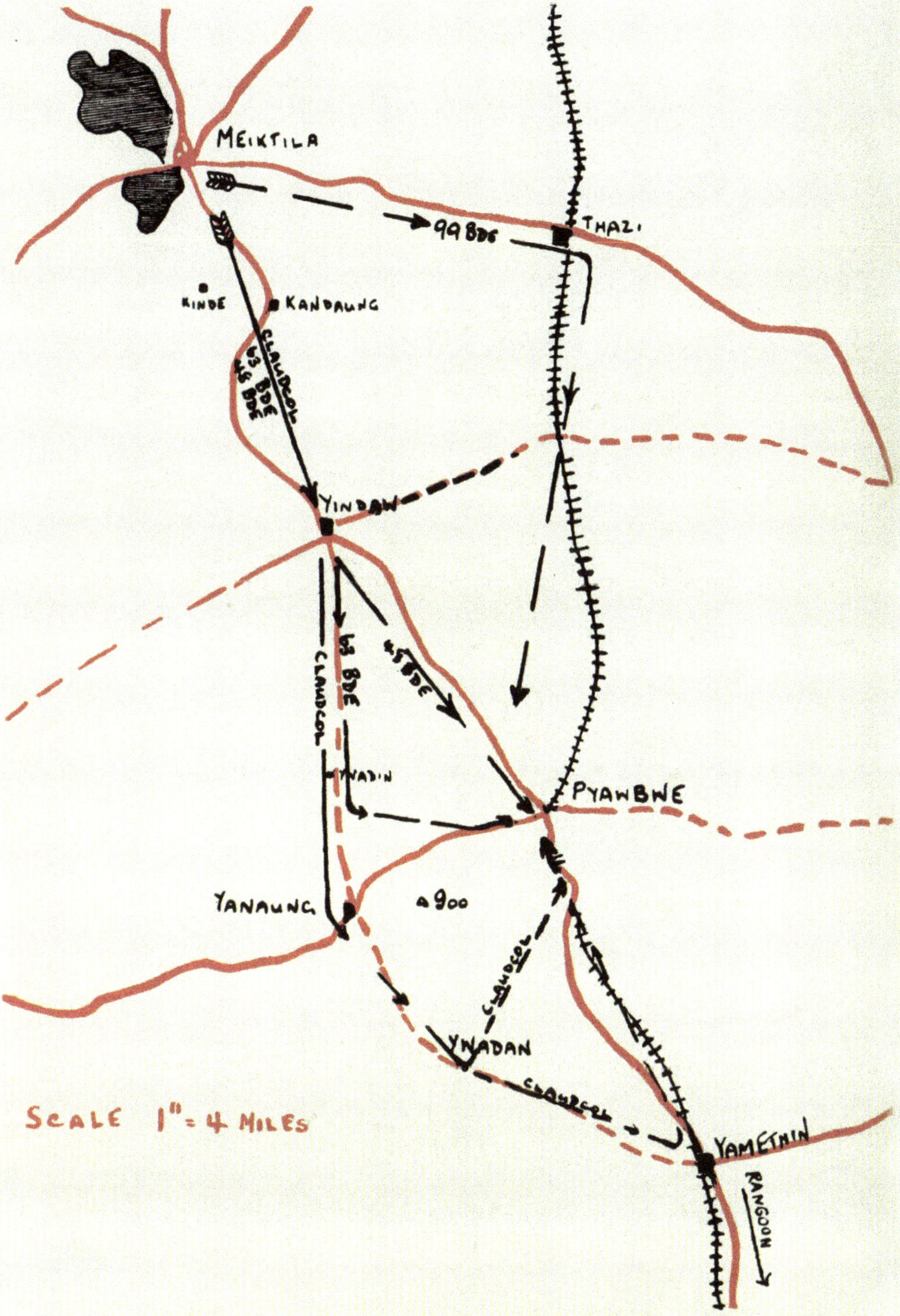

MEIKTILA
THAZI
99 BDE
KINDE
KANDAUNG
CLAUDCOL
63 BDE
48 BDE
YINDAW
CLAUDCOL
63 BDE
48 BDE
YWADIN
PYAWBWE
YANAUNG
900
CLAUDCOL
YWADAN
CLAUDCOL
YAMETHIN
RANGOON

SCALE 1" = 4 MILES

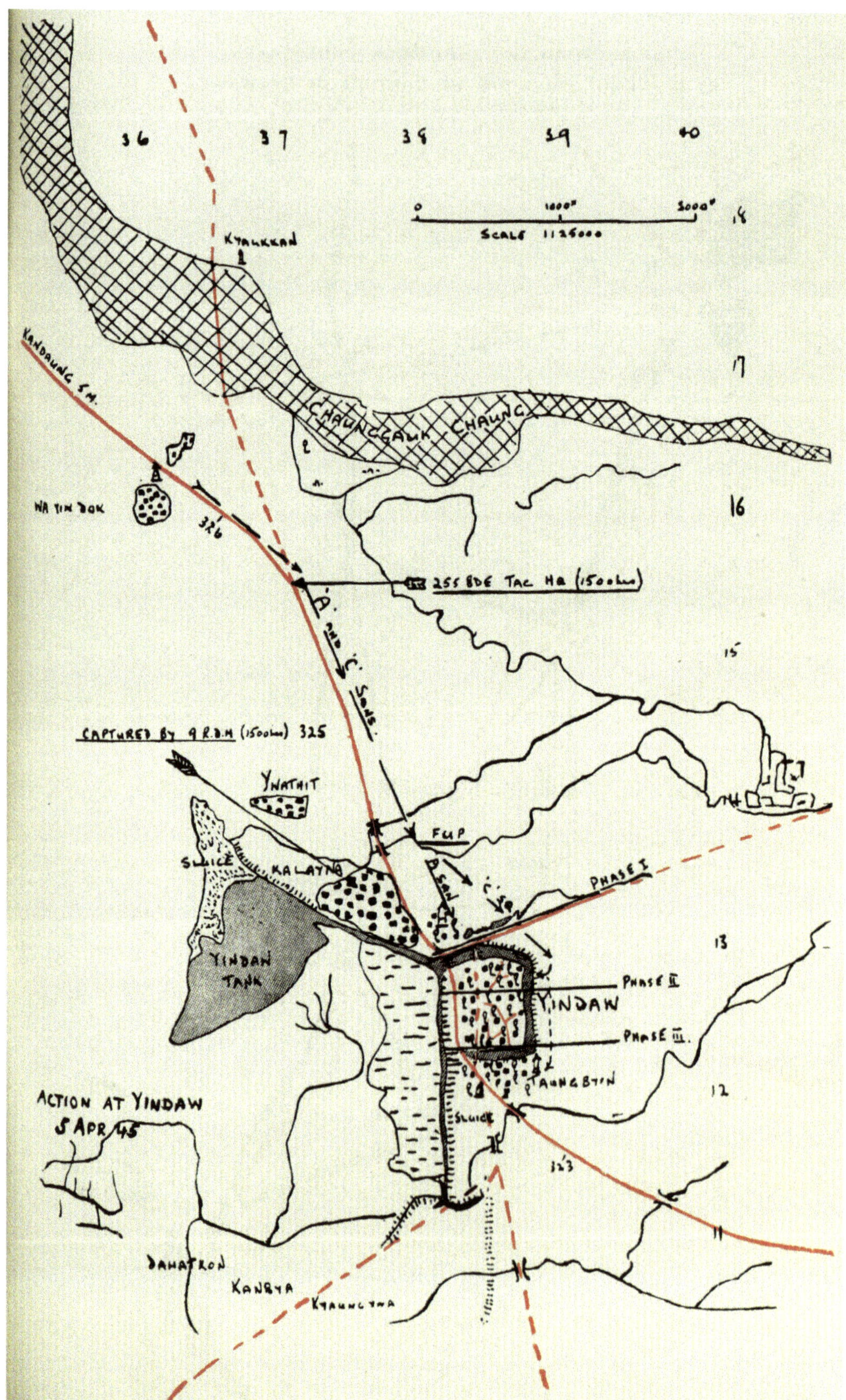
36
37
38
39
40
0
1000"
3000"
SCALE 1:25000
18
KYAUKKAN
KANDAUNG S.M.
17
CHAUNGGAUK CHAUNG
16
WA YIN DOK
326
255 BDE TAC HQ (1500hrs)
'A' AND 'C' SQDS.
15
CAPTURED BY 9 R.D.H (1500hrs) 325
YNATHIT
FUP
SLUICE
KALAYNA
PHASE I
13
YINDAW TANK
PHASE II
YINDAW
PHASE III
TAUNGBYIN
12
ACTION AT YINDAW
5 APR 45
SLUICE
123
11
DAHATKON
KANBYA
KYAUNGYNA

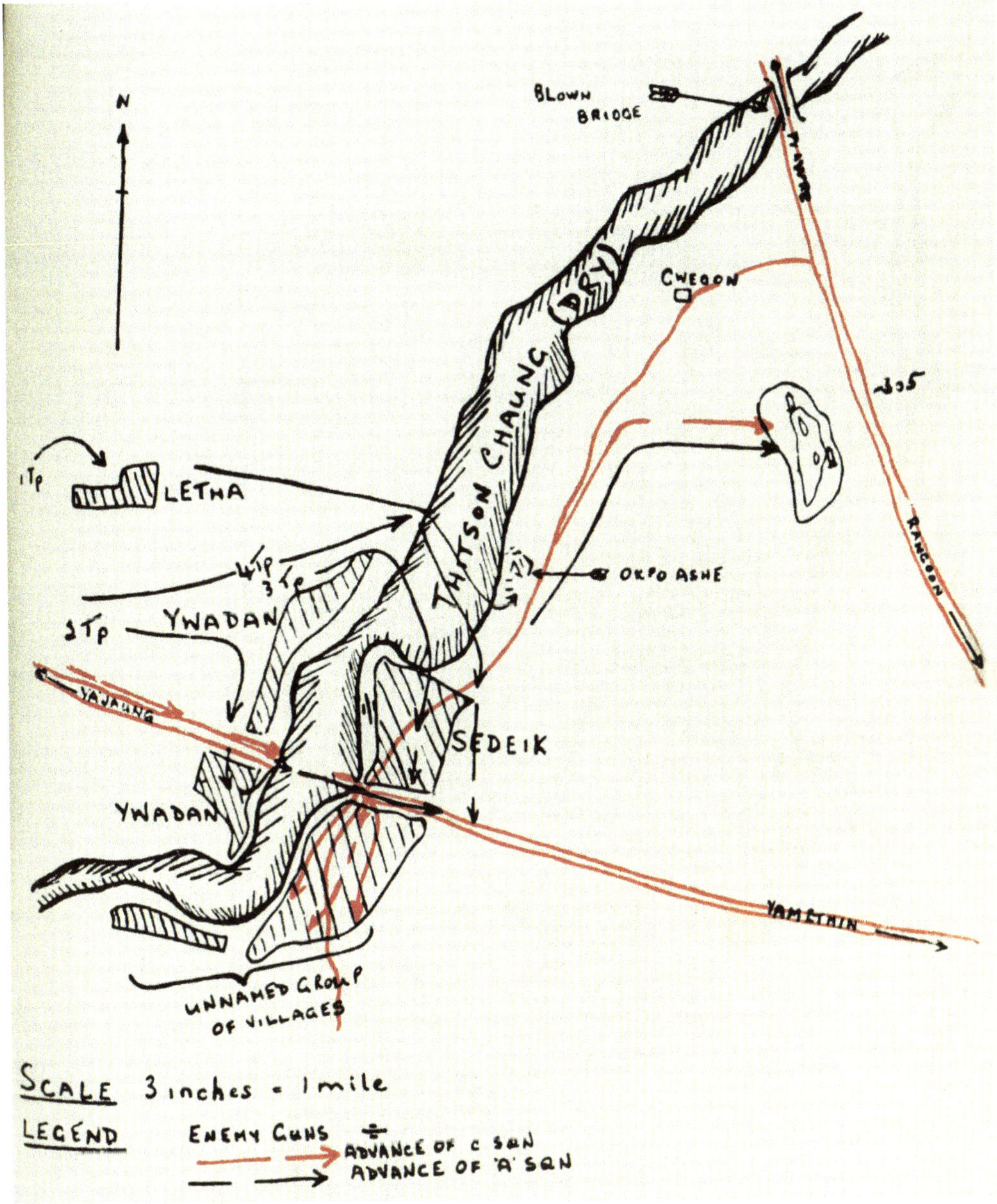

N
BLOWN BRIDGE
CHEGON
THITSON CHAUNG (DRY)
305
RANGOON
1Tp
LETHA
4Tp
3Tp
OKPO ASHE
2Tp
YWADAN
YAJAUNG
SEDEIK
YWADAN
YAMETHIN
UNNAMED GROUP OF VILLAGES
SCALE 3 inches = 1 mile
LEGEND
ENEMY GUNS
ADVANCE OF C SQN
ADVANCE OF 'A' SQN

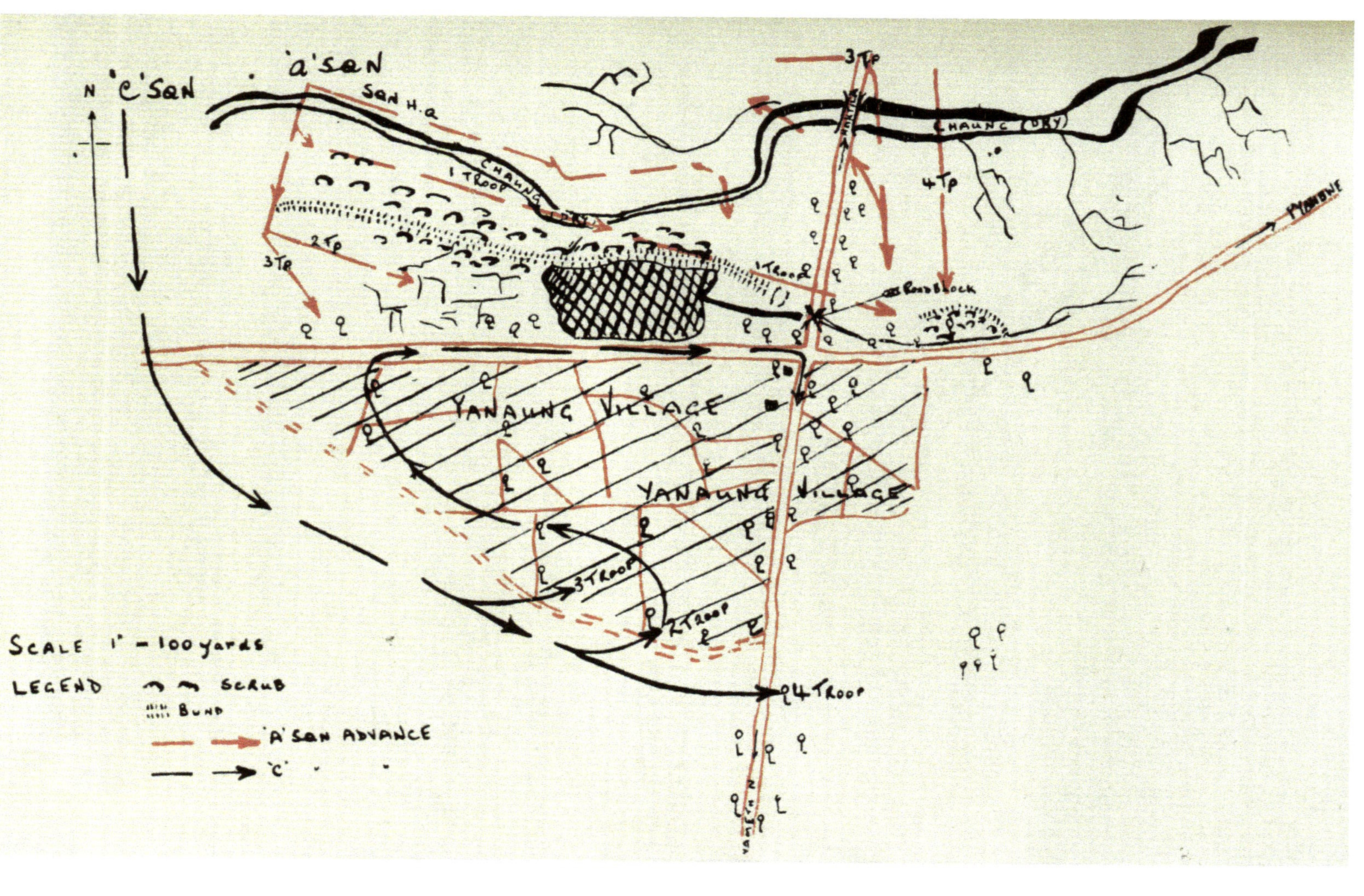

N
'C' SQN
'A' SQN
SQN H.Q
CHAUNG
1 TROOP
2 TP
3 TP
3 TP
CHAUNG (DRY)
4 TP
1 TROOP
ROAD BLOCK
YANAUNG VILLAGE
YANAUNG VILLAGE
3 TROOP
2 TROOP
4 TROOP
SCALE 1" - 100 yards
LEGEND
SCRUB
BUND
'A' SQN ADVANCE
'C' " "

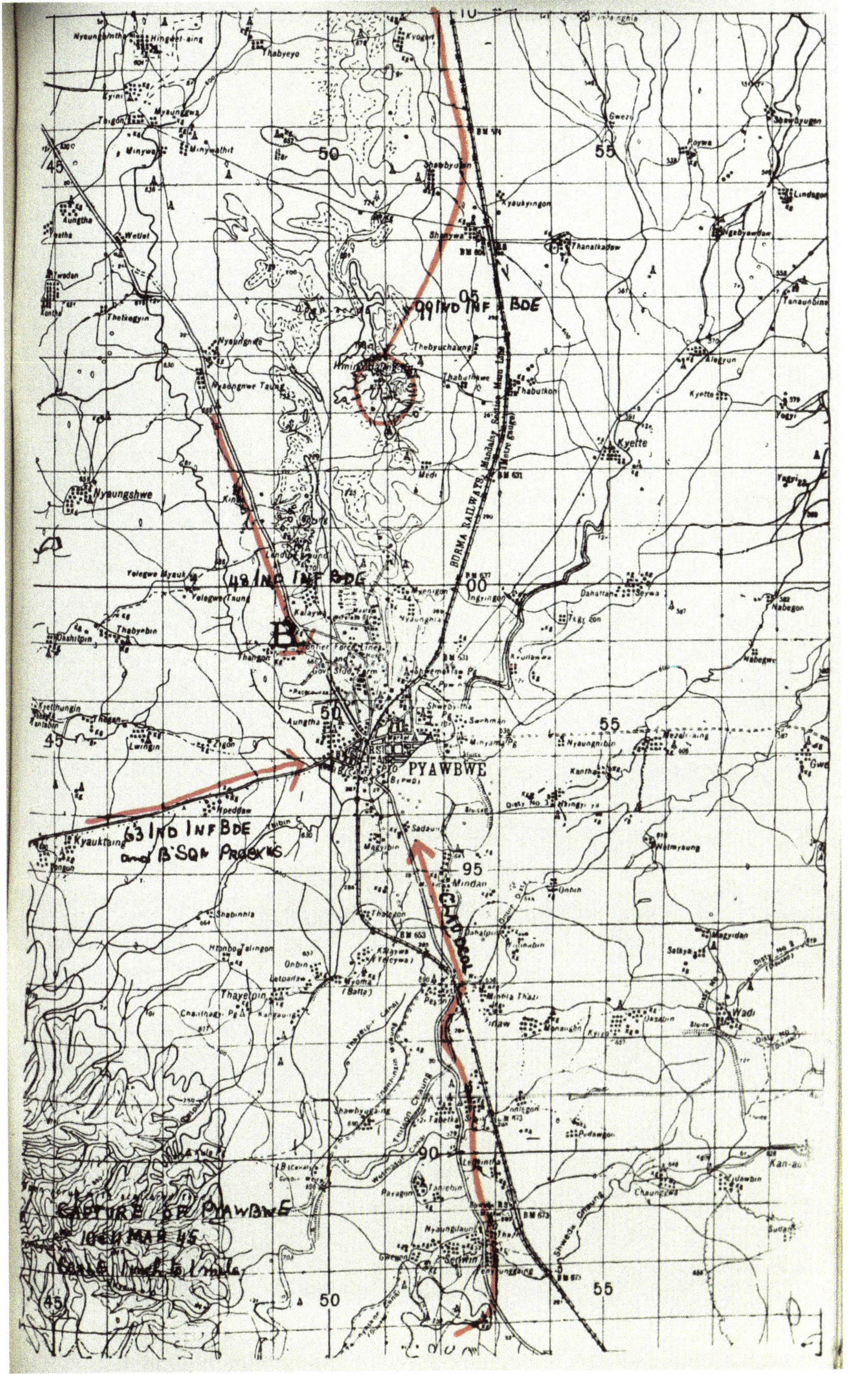
99 IND INF BDE
48 IND INF BDE
63 IND INF BDE
and 'B' SQN PROBYNS
CLAYPOOL
PYAWBWE
BURMA RAILWAYS
CAPTURE OF PYAWBWE
10th MAR 45
Scale 1 inch to 1 mile

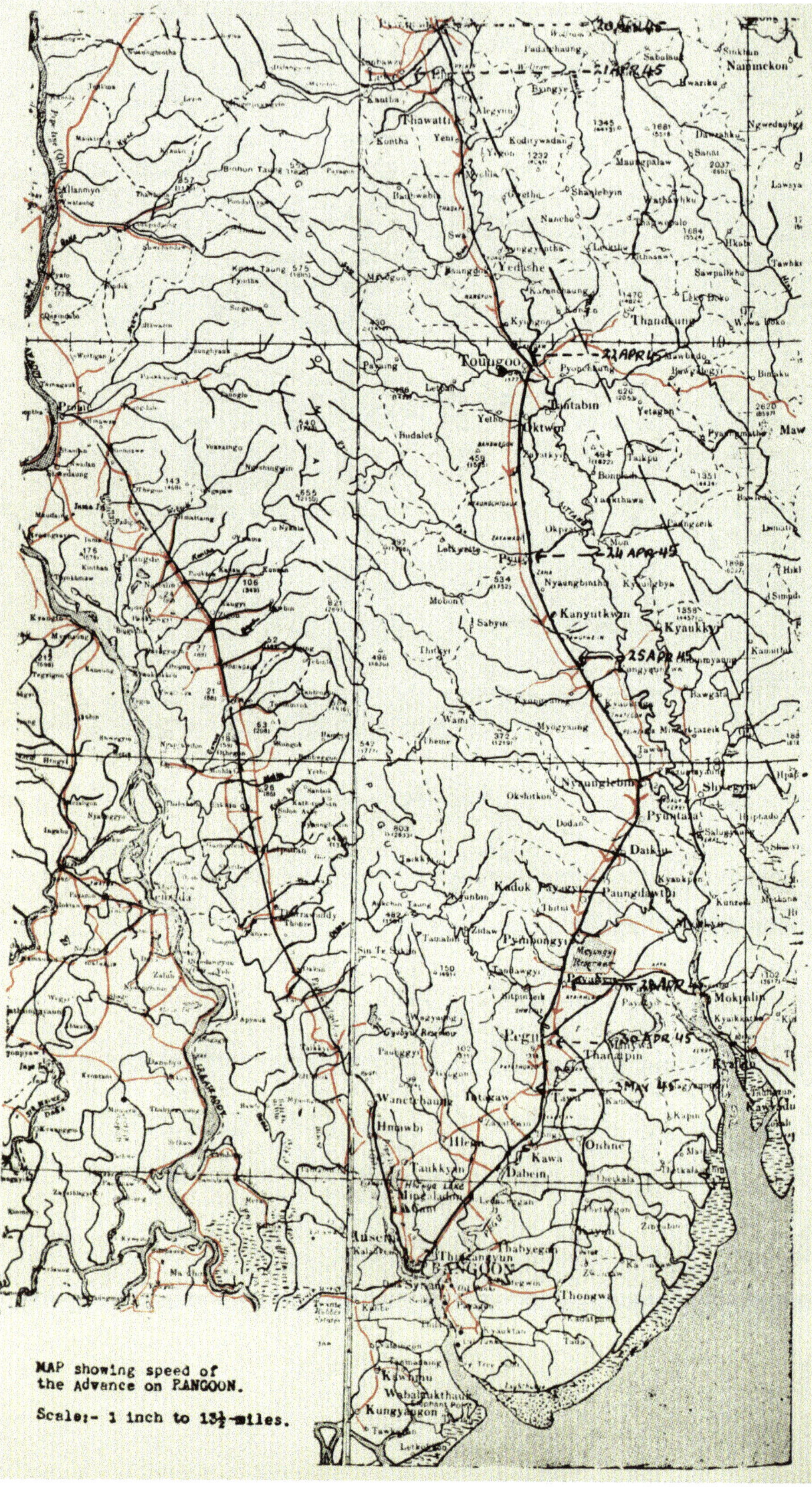

MAP showing speed of the Advance on RANGOON.

Scale:- 1 inch to 13½ miles.

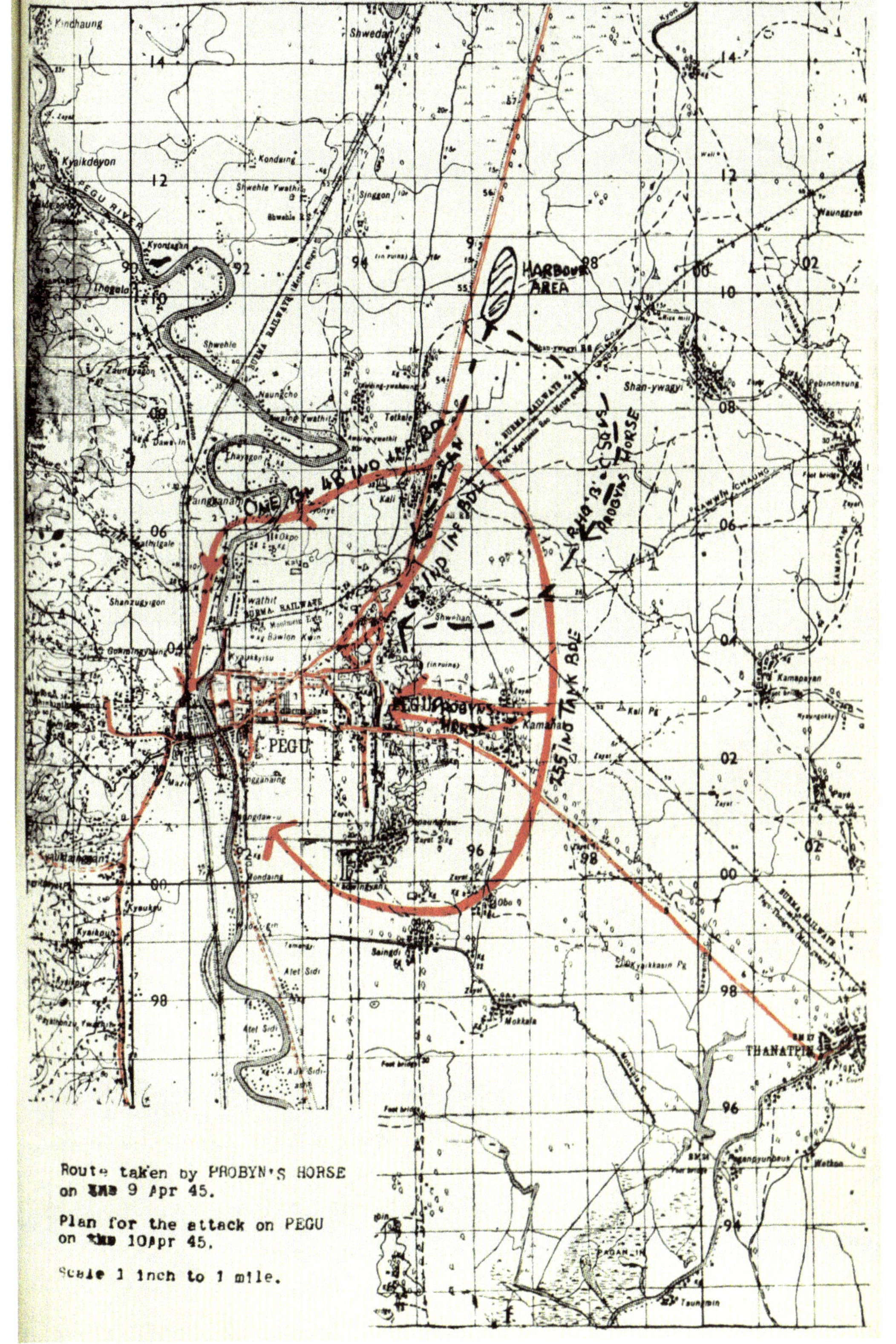

Route taken by PROBYN'S HORSE on 9 Apr 45.

Plan for the attack on PEGU on 10Apr 45.

Scale 1 inch to 1 mile.

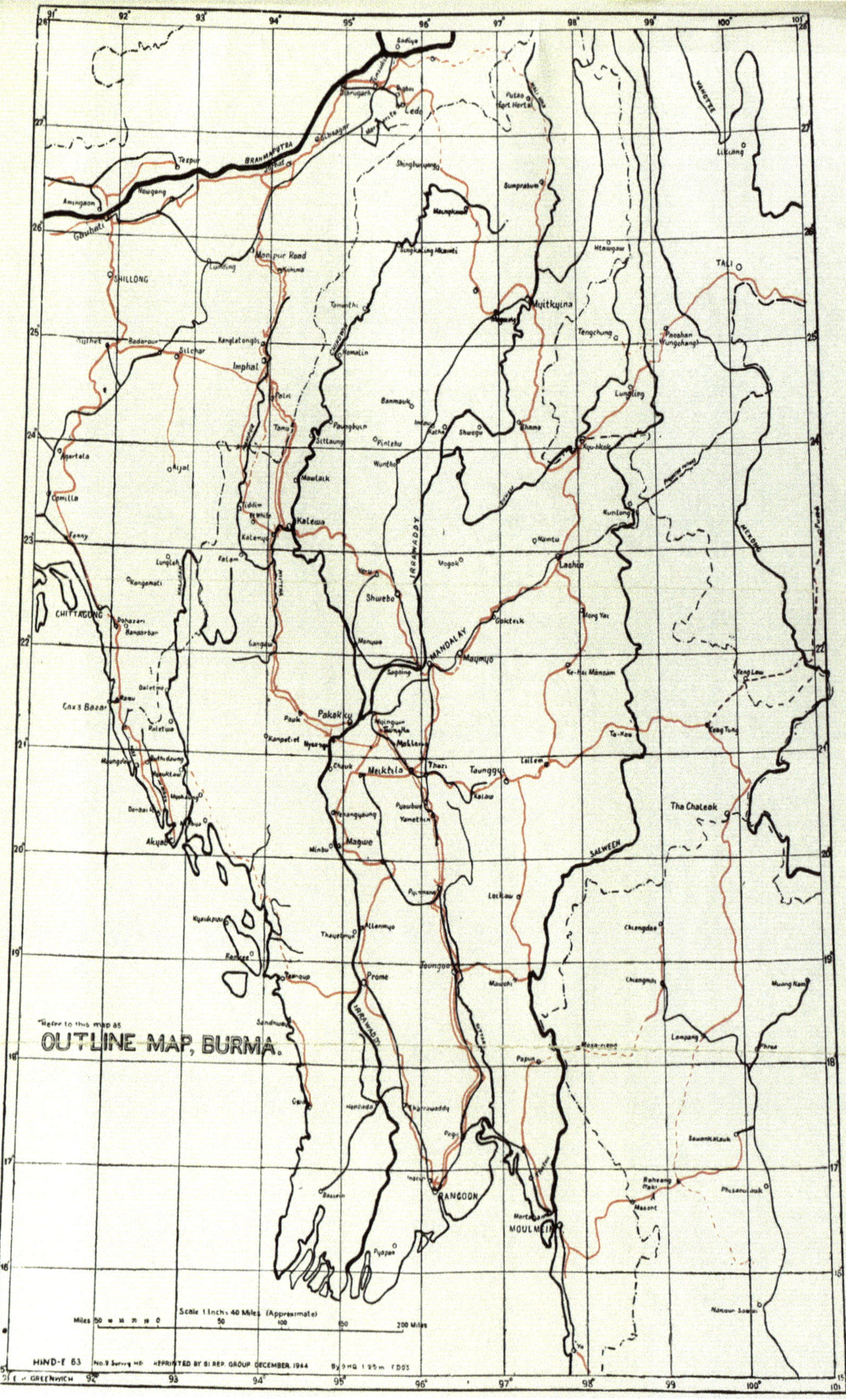
Refer to this map as
OUTLINE MAP, BURMA.
Scale 1 Inch = 40 Miles (Approximate)
HIND-E 63
REPRINTED BY 81 REP. GROUP DECEMBER 1944
E. of GREENWICH

www.ingramcontent.com/pod-product-compliance
Ingram Content Group UK Ltd.
Pitfield, Milton Keynes, MK11 3LW, UK
UKHW051030290726
14058UKWH00012B/859